My Book

This book belongs to

Name: _____

Copy right © 2019 MATH-KNOTS LLC

All rights reserved, no part of this publication may be reproduced, stored in any system or transmitted in any form, or by any means, electronic, mechanical, photocopying, recording, or otherwise without the written permission of MATH-KNOTS LLC.

Cover Design by :
Gowri Vemuri

First Edition :
May, 2020

Author :
Gowri Vemuri

Editor :
Ritvik Pothapragada

Questions: mathknots.help@gmail.com

Dedication

This book is dedicated to:
My Mom, who is my best critic, guide and supporter.
To what I am today, and what I am going to become tomorrow,
is all because of your blessings, unconditional affection and support.

This book is dedicated to the
strongest women of my life,
my dearest mom
and
to all those moms in this universe.

G.V.

GEOMETRY

INDEX

Notes	9 - 30
Identify the angle relationship	31 - 36
Find the angle measure	37 - 48
Find the value of x	49 - 80
Find the area of the 2-D figures	81 - 90
Find the circumference of a circle	91 - 94
Find the missing measure of the circle	95 - 97
Find the volume of 3-D figures	98 - 112
Find the total surface area of 3-D figures	113 - 128
Answer Keys	128 - 170

GEOMETRY

Notes

Geometry Notes

1. Complementary Angles:
Two angles that add up to 90 degrees are called as Complementary angles.

Example: $\angle X + \angle Y = 90$
$\angle X$ and $\angle Y$ are called complementary angles.

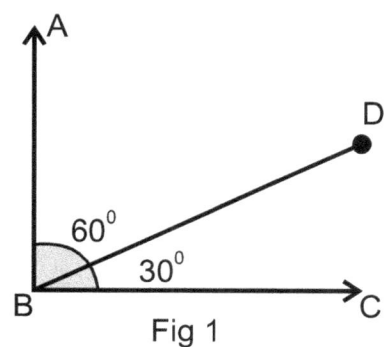
Fig 1

$30° + 60° = 90°$
$30°$ and $60°$ are complementary angles

2. Supplementary Angles:
Two angles that add up to 180 degrees are called as supplementary angles.
Example: $\angle a + \angle b = 180$
$\angle a$ and $\angle b$ are called complementary angles.

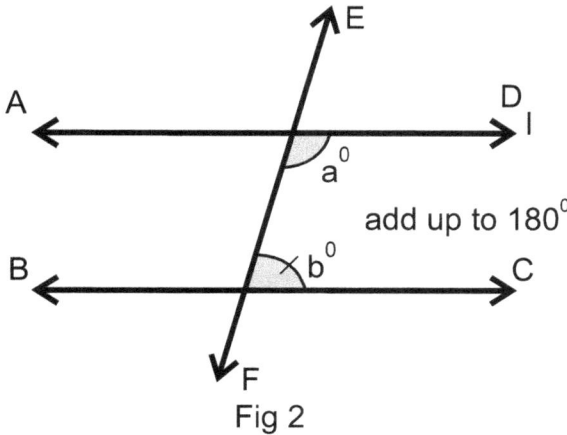
Fig 2

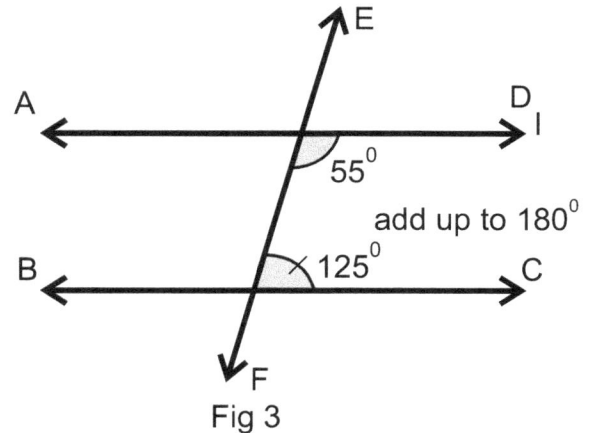
Fig 3

Fig 4

$\angle a + \angle b = 180 ; 55° + 125° = 180°$
$55°$ and $125°$ are called as complementary angles

$\angle a + \angle b = 180 ; 40° + 140° = 180°$
$40°$ and $140°$ are called as complementary angles

3. Vertical Angles:

Vertical angles are pairs of opposite angles made by intersecting lines.
If two angles are vertical, then they are congruent.
Example:
∠a and ∠b are called vertical angles
∠a = ∠b

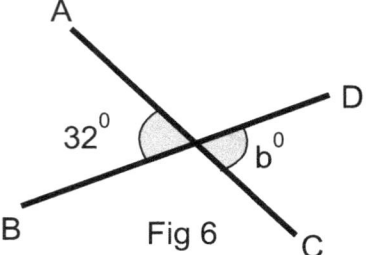

∠a and ∠b are called vertical angles and vertical angles are equal to each other.
In Fig 6 based on the rule of the vertical angles ∠b = 32^0

4. Adjacent Angles:

Two angles that have a common side and a common vertex (corner point), and don't overlap are called adjacent angles

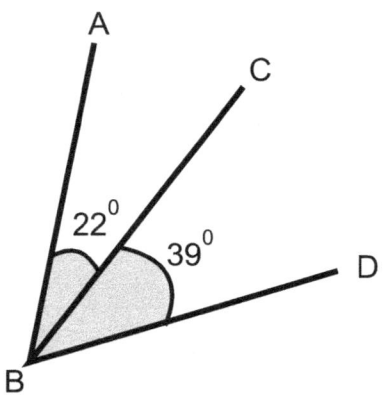

∠ABC and ∠CBD are called as adjacent angles as they share same vertex B

GEOMETRY

5. Corresponding Angles

When two lines are crossed by another line (Transversal), the angles in matching corners are called as corresponding angles. A pair of angles each of which is on the same side of one of two lines cut by a transversal and on the same side of the transversal

The angles which occupy the same relative position at each intersection where a straight line crosses two others. If the two lines are parallel, the corresponding angles are equal.

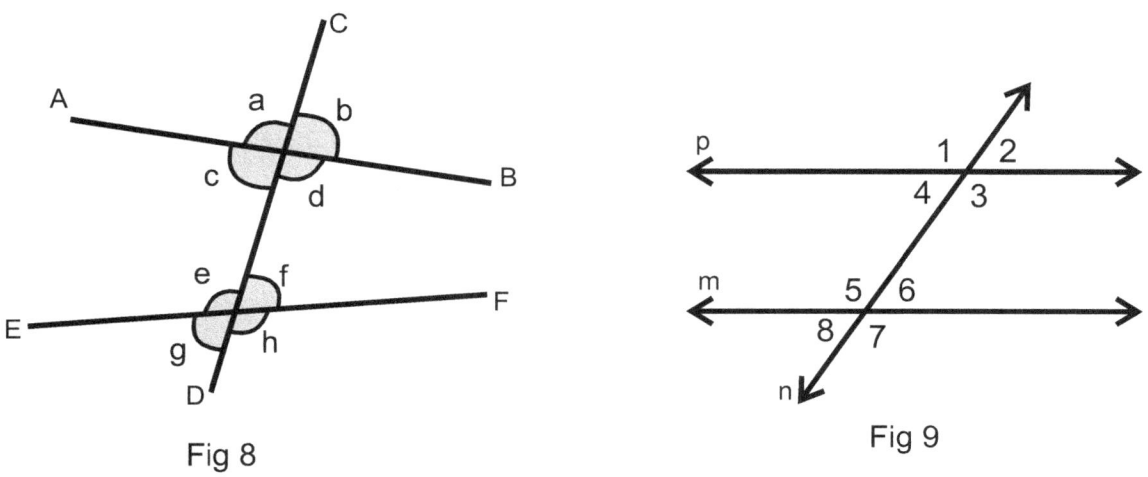

Fig 8   Fig 9

In Fig 8, $\overline{AB}$ and $\overline{EF}$ **are not parallel**, the corresponding angles ∠a and ∠e ; ∠c and ∠g ; ∠b and ∠f ; ∠d and ∠h are **not equal**.

In Fig 9, Lines p and m **are parallel**, the corresponding angles ∠1 and ∠5 ; ∠4 and ∠8 ; ∠2 and ∠6 ; ∠3 and ∠7 **are equal**.

6. Alternate Angles

Two angles, not adjoining one another, that are formed on opposite sides of a line that intersects two other lines. If the original two lines are parallel, the alternate angles are equal.
one of a pair of angles with different vertices and on opposite sides of a transversal at its intersection with two other lines:

1. Alternate Interior Angles are a pair of angles on the inner side (inside) of each of those two intersected lines but on opposite sides of the transversal. If two parallel lines are cut by a transversal, the alternate interior angles are congruent Examples of Alternate Interior Angles In the figure shown, l is the transversal that cut the pair of lines. Angles 3 and 4 and angles 1 and 2 are alternate interior angles.

GEOMETRY

Notes

2: **Alternate Exterior Angles** are a pair of angles on the outer side (outside) of each of those two intersected lines but on opposite sides of the transversal.

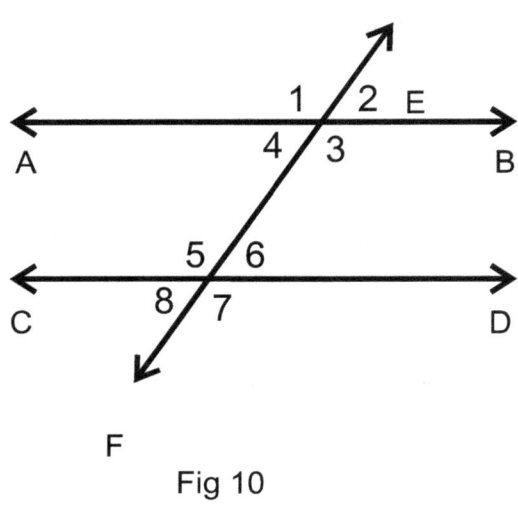

Fig 10

∠3, ∠4, ∠5, ∠6   Interior Angles
∠1, ∠2, ∠7, ∠8   Exterior Angles

∠4, ∠5 ⎫
∠3, ∠6 ⎭ Alternate Interior Angles

∠1, ∠8 ⎫
∠2, ∠7 ⎭ Alternate Exterior Angles

∠1, ∠5 ⎫
∠2, ∠6 ⎪
∠3, ∠7 ⎬ Corresponding Angles
∠4, ∠8 ⎭

7. Triangles:
Sum of the angles in any triangle are equal to 180 degrees.

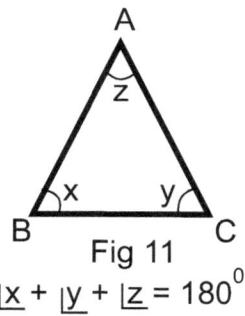
Fig 11
$\angle x + \angle y + \angle z = 180^0$

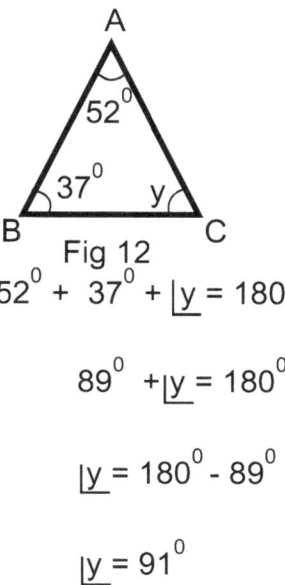
Fig 12
$52^0 + 37^0 + \angle y = 180^0$

$89^0 + \angle y = 180^0$

$\angle y = 180^0 - 89^0$

$\angle y = 91^0$

8. Quadrilaterals: Sum of the angles in any quadrilateral are equal to 360 degrees.

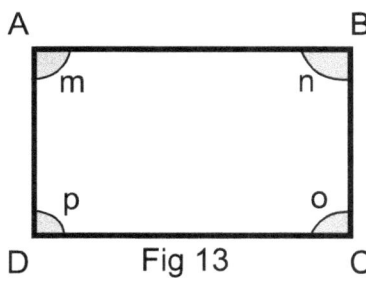
Fig 13

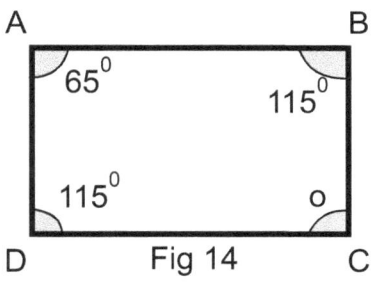
Fig 14

$\angle m + \angle n + \angle o + \angle p = 180^0$

$\angle m + \angle n + \angle o + \angle p = 180^0$

$115^0 + 115^0 + 65^0 + \angle p = 180^0$

$295^0 + \angle p = 180^0$

$\angle p = 180^0 - 295^0$

$\angle p = 65^0$

9. Area of a triangle :

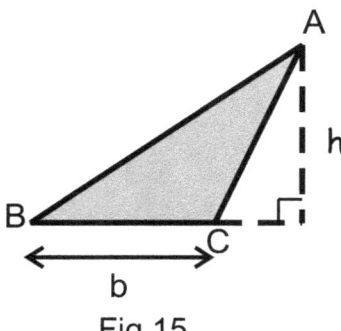
Fig 15

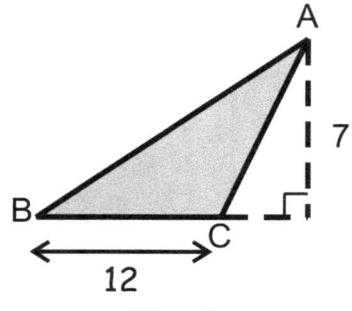
Fig 16

$A = \frac{1}{2} b h$

A = Area of the triangle
b = length of the base
h = height

$A = \frac{1}{2} b h$

$A = \frac{1}{2} \times (12)(7)$

$= (6)(7)$

$= 42$ sq. units

10. Perimeter of a triangle :

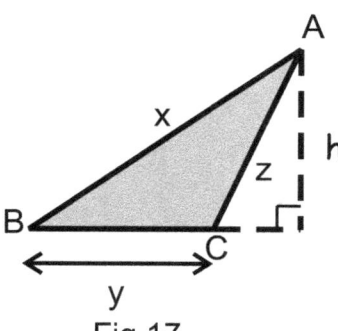
Fig 17

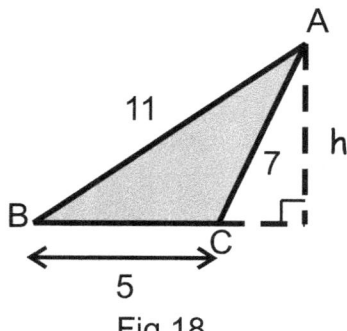
Fig 18

P = x + y + z
P = Perimeter
x, y, z are lengths of
the sides of the triangle

P = x + y + z
P = 11 + 5 + 7
P = 23 units

11. Perimeter and Area of a Square

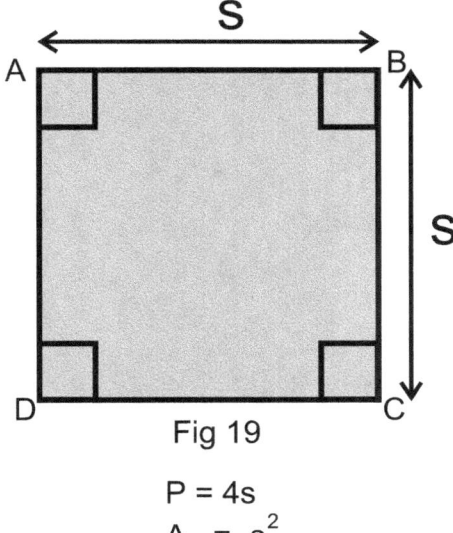

Fig 19

P = 4s
A = s^2

P = Perimeter
A = Area
s = Length of the side of the square

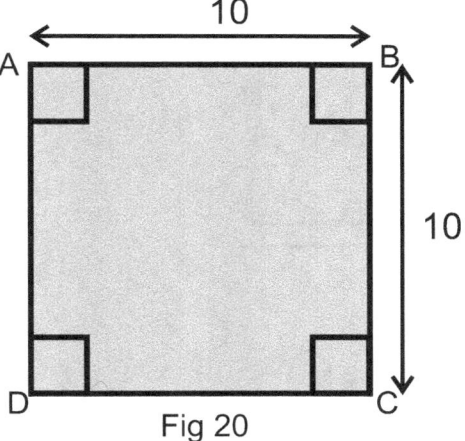

Fig 20

P = 4s
A = s^2

P = 4s
P = 4 X 10
P = 40 units

A = s^2
A = 10^2
A = 100 sq.units

12. Perimeter and Area of a Rectangle

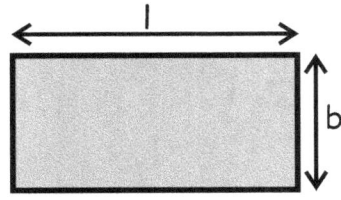

Fig 21

A = l X b

P = 2(l + b)
A = Area
P = Perimeter
l = length of the rectangle
b = width of the rectangle

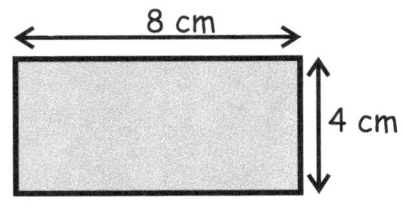

Fig 22

A = 8 X 4
A = 32 sq.cm

P = 2(8 + 4)
P = 2(12)
P = 24 cm

13. Area of a Trapezium

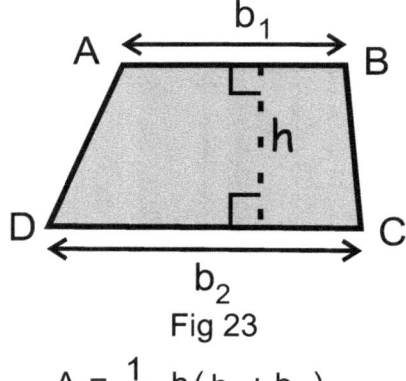

Fig 23

$A = \frac{1}{2} h(b_1 + b_2)$

A = Area
b_1, b_2 are the lengths of parallel sides
h = Distance between the parallel sides

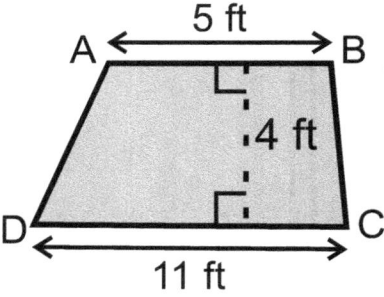

Fig 24

$A = \frac{1}{2} h(b_1 + b_2)$

$A = \frac{1}{2} 4(11 + 5)$

$A = 2(16)$

$A = 32$ sq.ft

14. Perimeter of a Trapezium

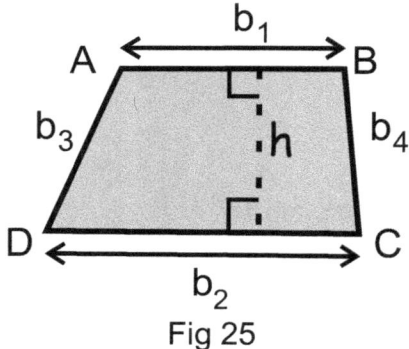

Fig 25

$P = b_1 + b_2 + b_3 + b_4$

P = Perimeter
b_1, b_2 are the lengths of parallel sides
b_3, b_4 are the lengths of non parallel sides

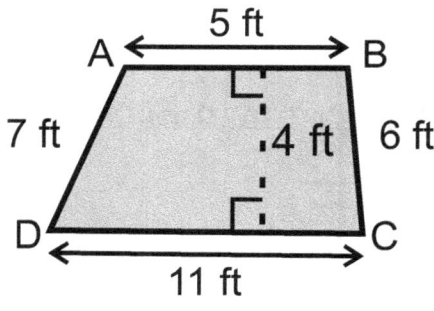

Fig 26

$P = b_1 + b_2 + b_3 + b_4$

$P = 5 + 11 + 7 + 6$

$P = 29$ ft

15. Area of a parellelogram

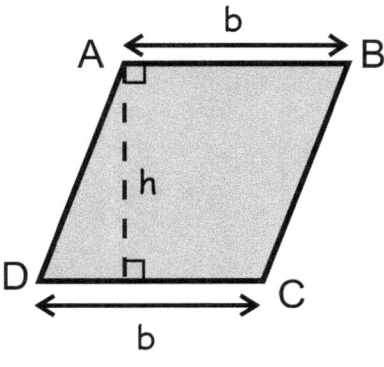
Fig 27

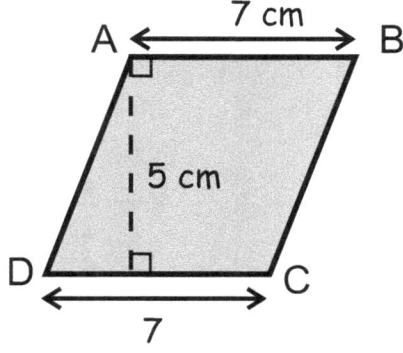
Fig 28

A = bh

A = Area

b = base

h = Height

A = bh

A = 5 X 7
A = 45 sq. cm

16. Perimeter of a parellelogram

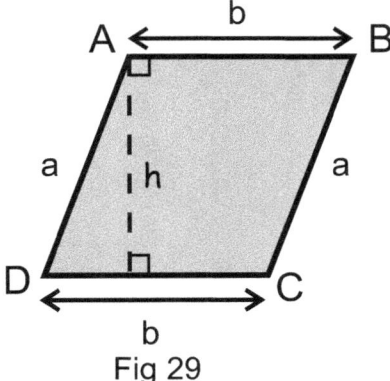
Fig 29

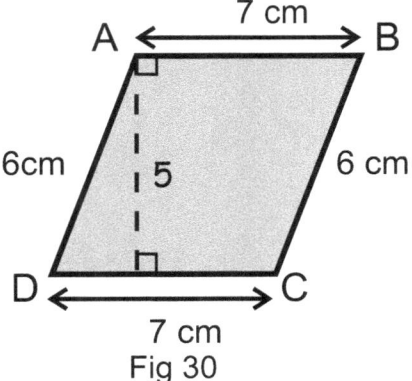
Fig 30

P = a + a + b + b
  = 2(a + b)

P = 6 + 6 + 7 + 7
  = 2(6 + 7)
  = 2(13)
  = 26 cm

17. Circumference and Area of a Circle

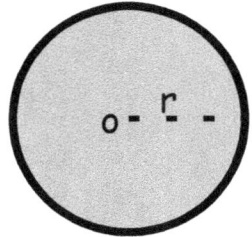

Fig 31

C = Πr

A = Πr²

pi

Π = 3.14

Π = $\frac{22}{7}$

C = Circumference of the circle

A = Area of the circle
r = radius

Note : Diameter(d) = 2r

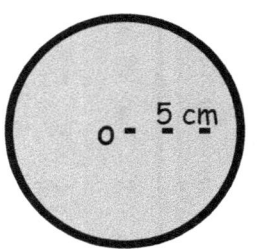

Fig 32

C = 2Πr

C = 2 X 3.14 X 5

C = 10 X 3.14

C = 31.4 cm

A = 3.14 X 5 X 5
A = 3.14 X 25
A = 78.5 sq.cm

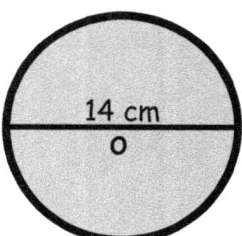

Fig 33

Diameter(d) = 14 cm

Radius = $\frac{14}{2}$

Radius = 7 cm

C = 2 X 3.14 X 7

C = 14 X 3.14

C = 43.86cm

A = 3.14 X 7 X 7
A = 3.14 X 49
A = 153.86 sq.cm

18. Volume of a sphere

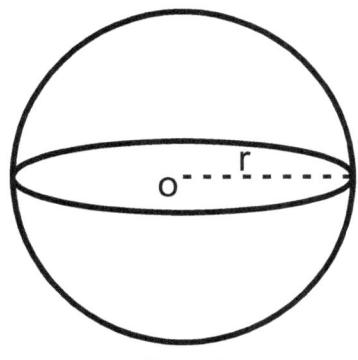

Fig 34

$$V = \frac{4}{3} \Pi r^3$$

V = Volume of the sphere
r = Radius of the sphere

Diameter = Twice the radius

pi
$\Pi = 3.14$
$\Pi = \frac{22}{7}$

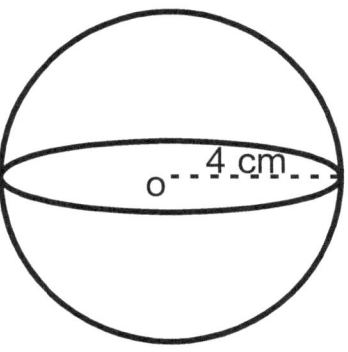

Fig 35

$$V = \frac{4}{3} \Pi r^3$$

$$V = \frac{4}{3} \times 3.14 \times 4^3$$

$$V = 268.08 \text{ cm}^3$$

19. Surface area of a sphere

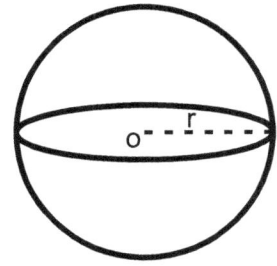

Fig 36

$S = 4\Pi r^2$

pi
$\Pi = 3.14$
$\Pi = \frac{22}{7}$

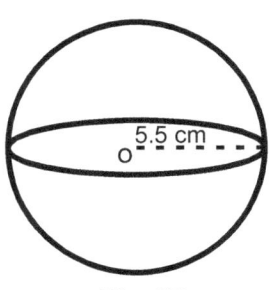

Fig 37

$S = 4\Pi r^2$

$S = 4 \times 3.14 \times (5.5)^2$

$S = 379.94 \text{ cm}^2$

20. Volume of a cone

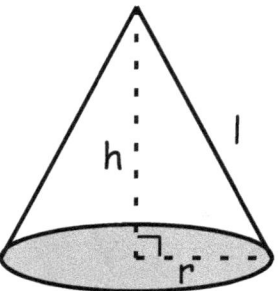

Fig 38

$$V = \frac{1}{3} \Pi r^2 h$$

V = Volume of the cone
r = Radius
h = Height
l = Slant height

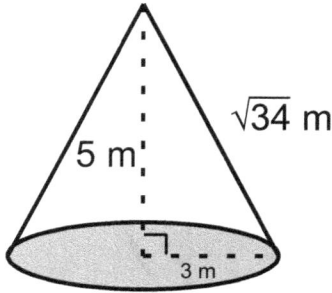

Fig 39

$$V = \frac{1}{3} \Pi r^2 h$$

$$V = \frac{1}{3} \times 3.14 \times 3^2 \times 5$$

$$V = 47.1239 \text{ m}^3$$

21. Lateral surface area of a cone

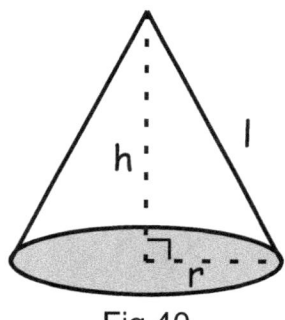

Fig 40

$$\begin{aligned} \text{L.S.A} &= \Pi r l \\ &= \Pi r \times \sqrt{r^2 + h^2} \end{aligned}$$

$$l = \sqrt{r^2 + h^2}$$

L.S.A = Lateral surface area
l = Slant height of the cone
r = radius
h = height

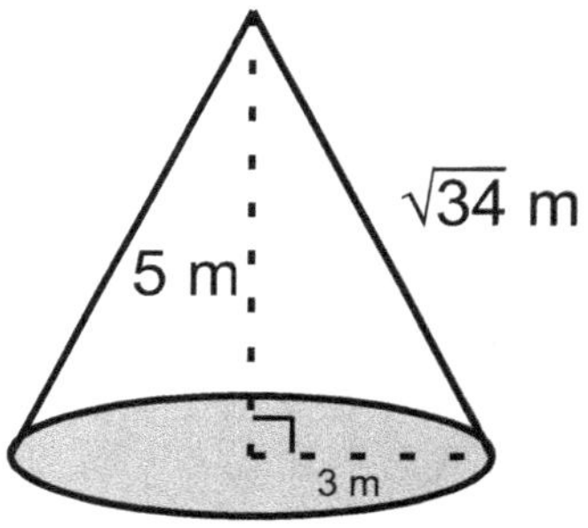

Fig 41

$$\text{L.S.A} = \Pi r l$$

$$\text{L.S.A} = 3.14 \times 3 \times \sqrt{34}$$

$$= 54.9554 \text{ m}^2$$

GEOMETRY

22. Base surface area of a cone

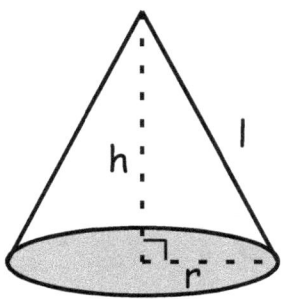

Fig 42

B.S.A = Πr^2

B.S.A = Base surface area
r = radius
h = height

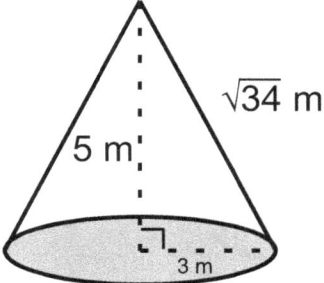

Fig 43

B.S.A = Πr^2

B.S.A = 3.14 X 3^2

B.S.A = 28.2743 m^2

23. Total surface area of a cone

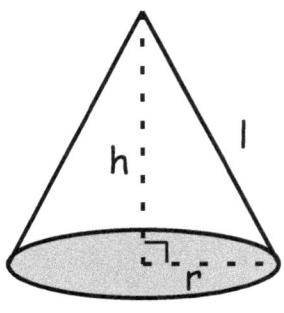

Fig 44

T.S.A = L.S.A + B.S.A

T.S.A = $\Pi rl + \Pi r^2$

T.S.A = $\Pi r(l + r)$

T.S.A = $\Pi r(r + \sqrt{r^2 + h^2})$

T.S.A = Total surface area
r = radius
h = height

Fig 45

T.S.A = $\Pi r(l + r)$

T.S.A = 3.14 X 3 ($\sqrt{34}$ + 3)

T.S.A = 83.2298 m^2

24. Volume of a Cylinder

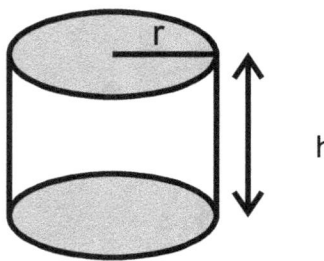

Fig 46

$V = \Pi r^2 h$

V = Volume
r = radius
h = height

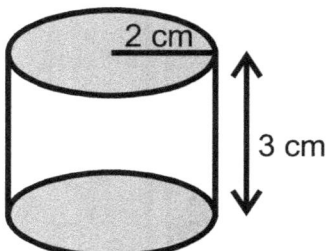

Fig 47

$V = \Pi r^2 h$

$V = 3.14 \times 2^2 \times 3$

$V = 37.6991 \text{ cm}^3$

25. Lateral Surface area of a Cylinder

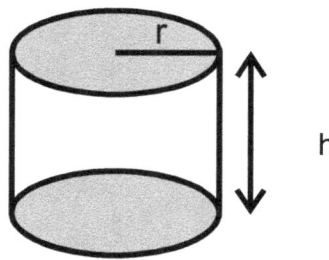

Fig 48

$L.S.A = 2 \Pi r h$

L.S.A = Lateral surface area
r = radius
h = height

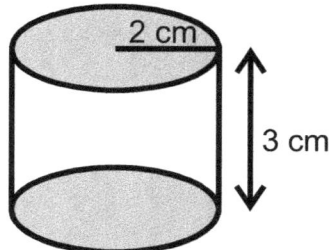

Fig 49

$L.S.A = 2 \Pi r h$

$L.S.A = 2 \times 3.14 \times 2 \times 3$

$L.S.A = 37.6991 \text{ cm}^2$

GEOMETRY

Notes

26. Top and bottom Surface area of a Cylinder

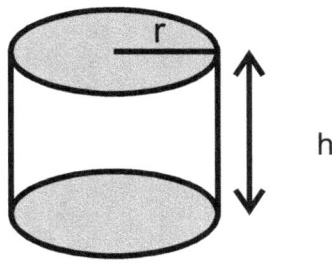

Fig 50

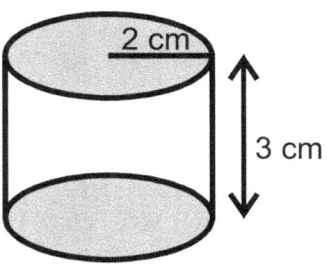

Fig 51

B.S.A = Πr^2

B.S.A = Bottom surface area
r = radius
h = height

B.S.A = Πr^2

B.S.A = 3.14 X 2^2

B.S.A = 12.5664 cm^2

Top and bottom surface area of the cylinder are equal. While calculating total surface area of the cylinder, remember to add the bottom surface are of the cylinder twice

---

T.S.A = Total surface area
T.S.A = L.S.A + B.S.A + B.S.A

T.S.A = $2\Pi rh + \Pi r^2 + \Pi r^2$

T.S.A = $2\Pi rh + 2\Pi r^2$

T.S.A = $2\Pi r (h + r)$

T.S.A = $2\Pi r (h + r)$

T.S.A = 2 X 3.14 X 2 (3 + 2)

T.S.A = 62.8318 cm^2

27. Volume of a Cuboid

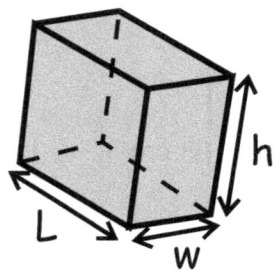

Fig 52

V = lwh
V = Volume
l = Length
w = Width or breadth
h = Height

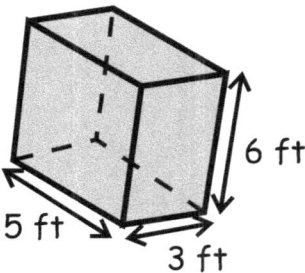

Fig 53

V = lwh

V = 5 X 3 X 6

V = 90 ft^3

28. Surface area of a Cuboid

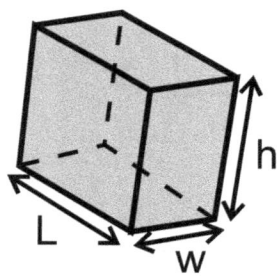

Fig 54

T.S.A = 2(lw + lh + wh)
T.S.A = Total surface area
l = Length
w = Width or breadth
h = Height

L.S.A = 2(lh + wh)
L.S.A = Lateral surface area
l = Length
w = Width or breadth
h = Height

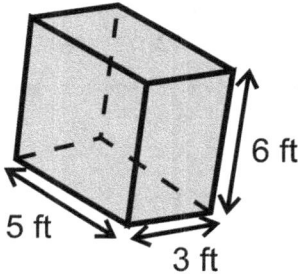

Fig 55

T.S.A = 2(lw + lh + wh)

T.S.A = 2(5X3 + 5X6 + 3X6)

T.S.A = 126 ft^2

L.S.A = 2(lh + wh)

L.S.A = 2(5X6 + 3X6)

L.S.A = 96 ft^2

29. Volume of a Cube

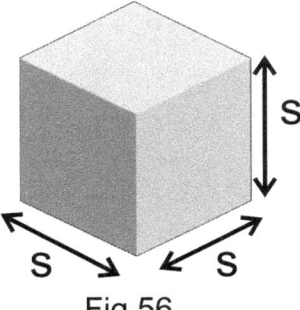
Fig 56

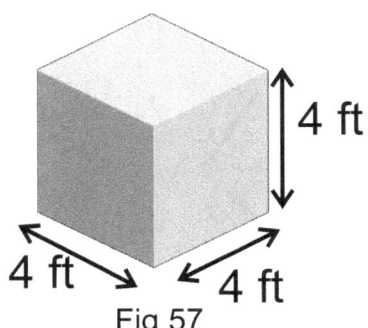
Fig 57

$V = s^3$
V = Volume
s = Side length of the cube

$V = s^3$
$V = 4^3$
$V = 64 \text{ ft}^3$

30. Surface area of a Cube

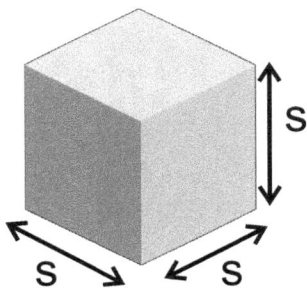
Fig 58

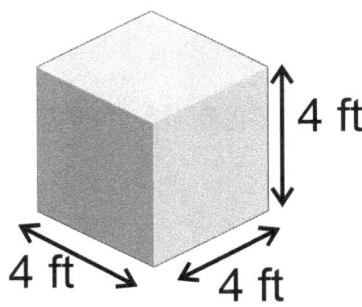
Fig 59

$T.S.A = 6s^2$
$T.S.A$ = Total surface area
s = Side length of the cube

$T.S.A = 6s^2$
$T.S.A = 6 \times 4^2$
$T.S.A = 96 \text{ ft}^2$

31. Volume of a square Pyramid

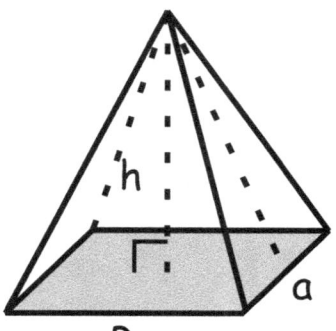

Fig 60

$V = \dfrac{1}{3} Bh$

V = Volume
B = Base area
h = Height

$V = \dfrac{1}{3} a^2 h$

V = Volume
a = side length
h = Height

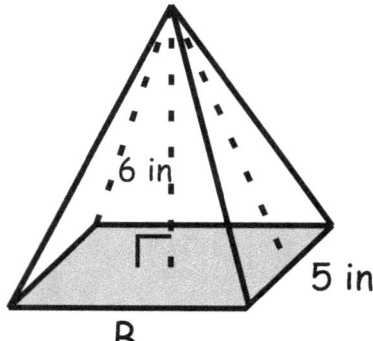

Fig 61

$V = \dfrac{1}{3} a^2 h$

$V = \dfrac{1}{3} \times 5^2 \times 6$

$V = \dfrac{1}{3} \times 150$

$V = 50 \text{ in}^3$

32. Lateral surface area of a square Pyramid

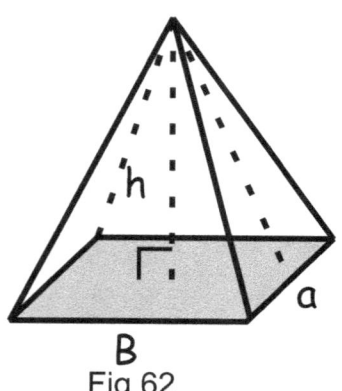

Fig 62

$S.A = a \times \sqrt{a^2 + 4h^2}$

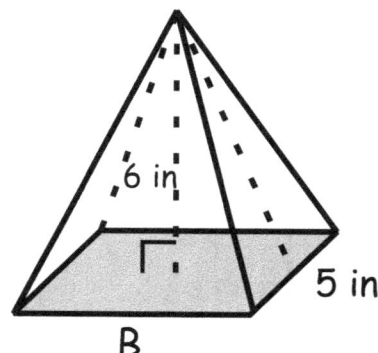

Fig 63

$S.A = a \times \sqrt{a^2 + 4h^2}$

$S.A = 5 \times \sqrt{5^2 + 4 \times 6^2}$

$S.A = 65 \text{ in}^2$

32. Base surface area of a square Pyramid

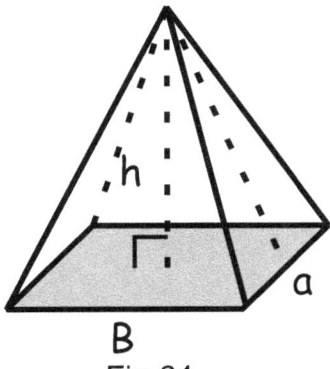
Fig 64

$B.S.A = a^2$

$T.S.A = a^2 + a \times \sqrt{a^2 + 4h^2}$

B.S.A = Base surface area
T.S.A = Total surface area
T.S.A = B.S.A + L.S.A

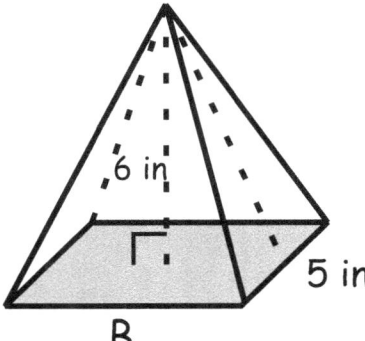
Fig 65

$B.S.A = 5 \times 5$

$B.S.A = 25 \text{ in}^2$

$T.S.A = 5^2 + 5 \times \sqrt{5^2 + 4 \times 6^2}$

$T.S.A = 25 + 65$

$T.S.A = 90 \text{ in}^2$

32. Volume of a Triangular prism

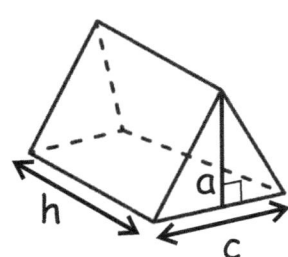
Fig 66

$V = Bh$

$V = \frac{1}{2} ach$

V = Volume
a = apothem
h = height
c = base length of the triangle

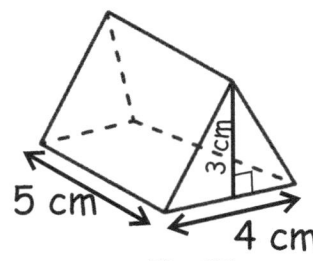
Fig 67

$V = Bh$

$V = \frac{1}{2} ach$

$V = \frac{1}{2}(3 \times 4 \times 5)$

$V = 30 \text{ cm}^3$

32. Volume of a triangular pyramid

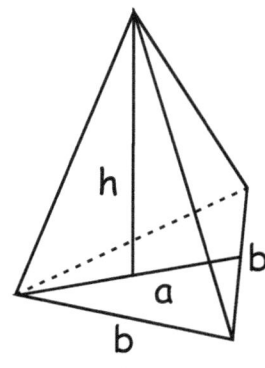

Fig 68

Volume of Triangular pyramid = $\frac{1}{6}$ abh

a = Apothem length of the pyramid
b = base length of the pyramid
h = Height of the pyramid

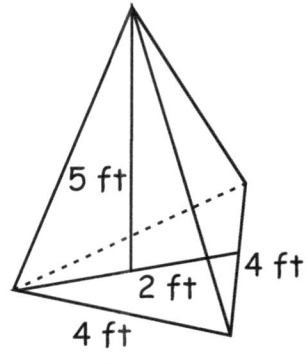

Fig 69

Volume of Triangular pyramid = $\frac{1}{6}$ abh

Volume of Triangular pyramid = $\frac{1}{6}$ (2X4X5)

Volume of Triangular pyramid = $\frac{1}{6}$ (40)

Volume of Triangular pyramid = 6.66 ft^3

33. Volume of a pentagonal pyramid

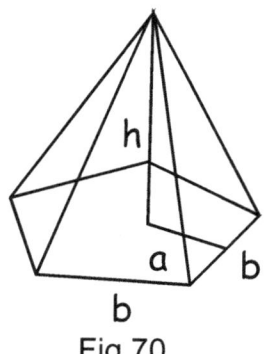

Fig 70

Volume of pentagonal pyramid = $\frac{5}{6}$ abh

a = Apothem length of the pyramid
b = base length of the pyramid
h = Height of the pyramid

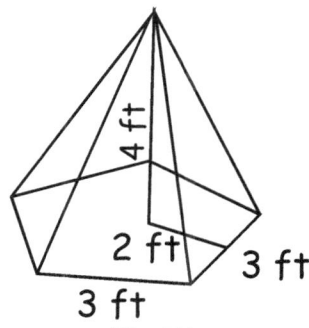

Fig 71

Volume of pentagonal pyramid = $\frac{5}{6}$ abh

Volume of pentagonal pyramid = $\frac{5}{6}$ (2X3X4)

Volume of pentagonal pyramid = $\frac{5}{6}$ (24)

Volume of pentagonal pyramid = 20 ft^3

34. Volume of a hexagonal pyramid

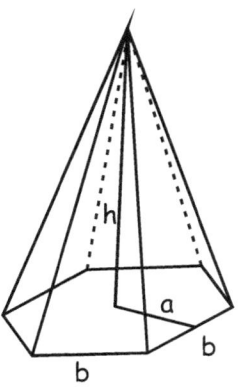

Fig 72

Volume of Hexagonal pyramid = abh

a = Apothem length of the pyramid
b = base length of the pyramid
h = Height of the pyramid

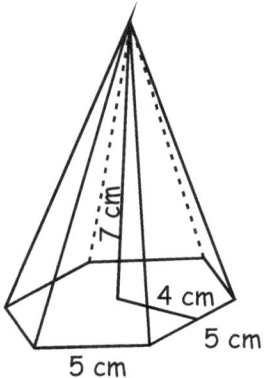

Fig 73

Volume of Hexagonal pyramid = 5X4X7
Volume of Hexagonal pyramid = 140 cm^3

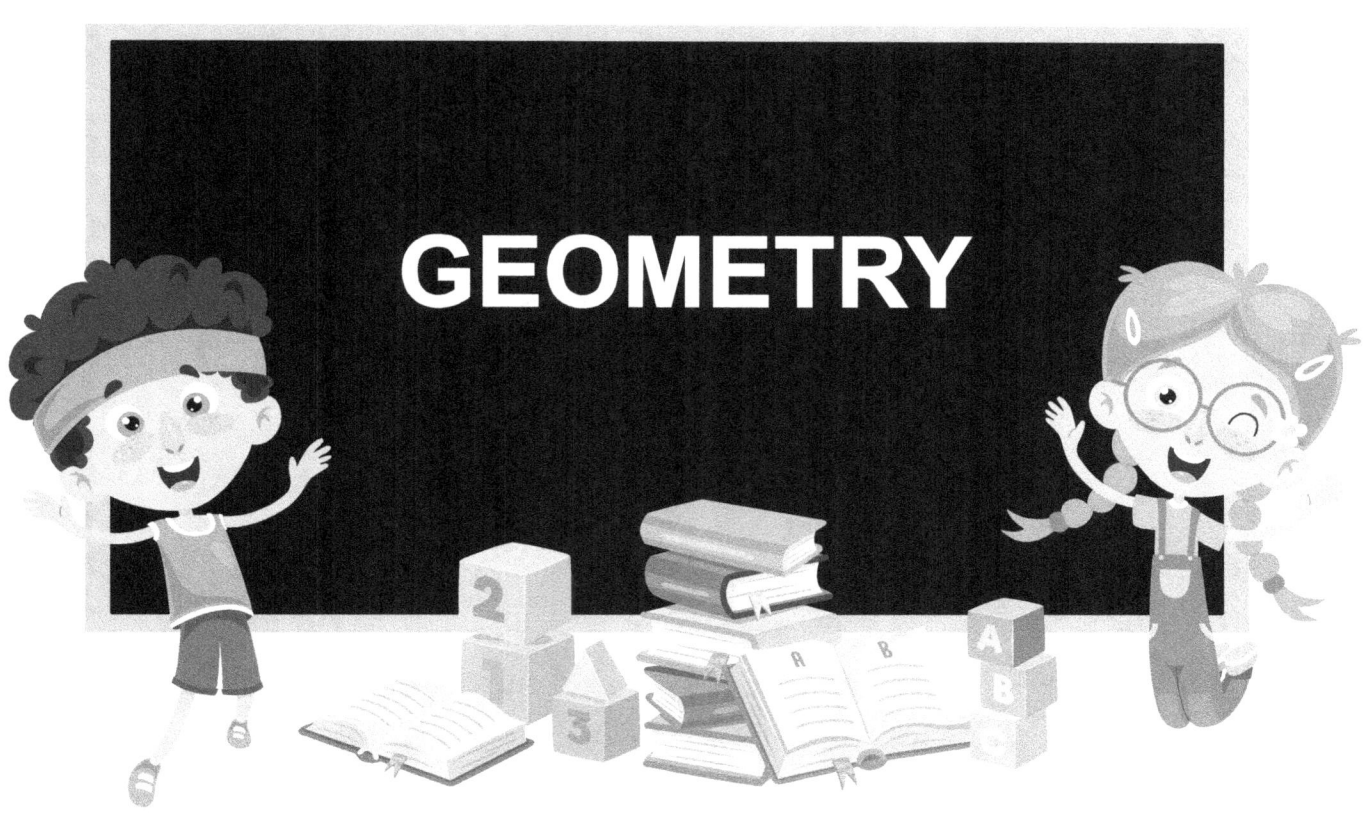

Identify the angles relationship from the below figures. Label them as complementary, supplementary, vertical, adjacent, alternate interior, corresponding, or alternate exterior angles.

(1)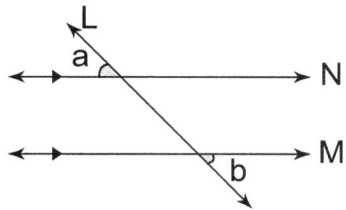

(2)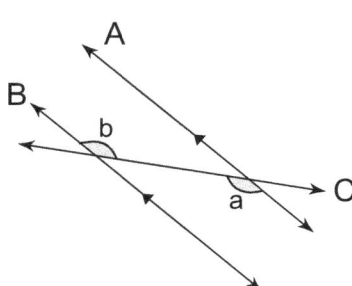

(3)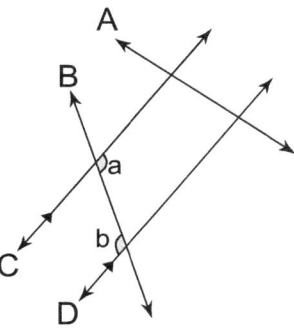

(4)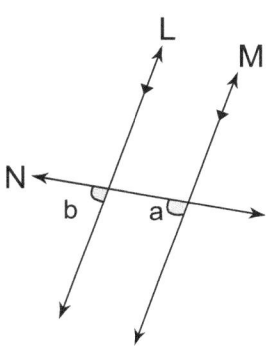

(5)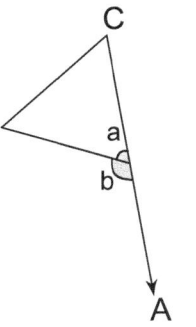

(6)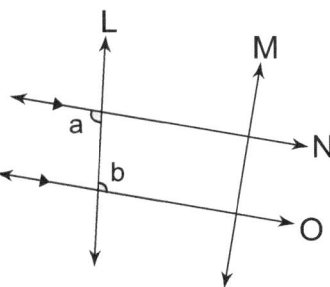

(7)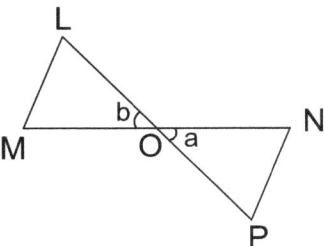

(8)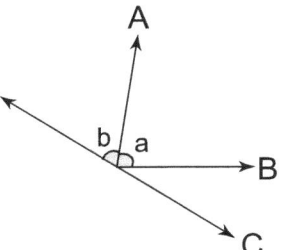

 **GEOMETRY**

Basic Math

Identify the angles relationship from the below figures. Label them as complementary, supplementary, vertical, adjacent, alternate interior, corresponding, or alternate exterior angles.

(9)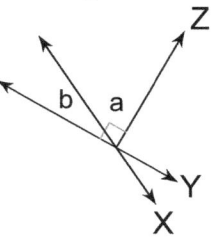

(10)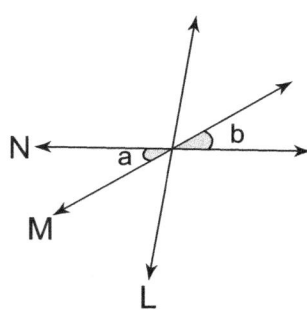

(11)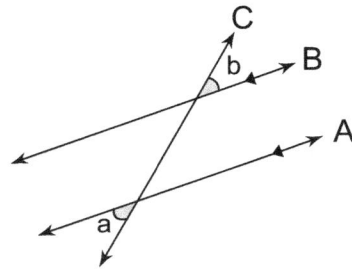

(12)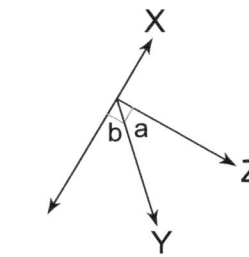

(13)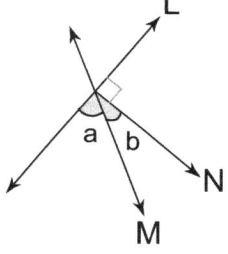

(14) 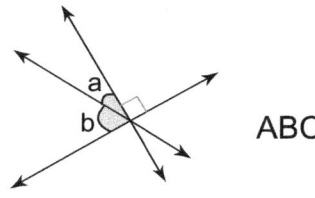   ABC

(15)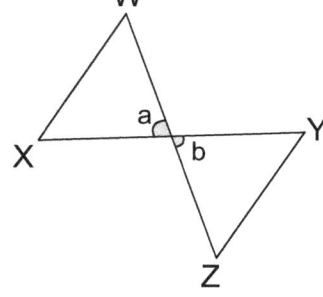

(16)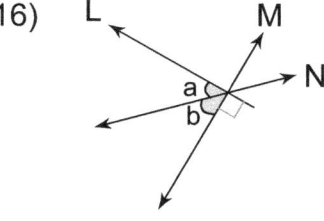

Identify the angles relationship from the below figures. Label them as complementary, supplementary, vertical, adjacent, alternate interior, corresponding, or alternate exterior angles.

(17)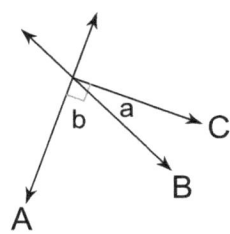

(18)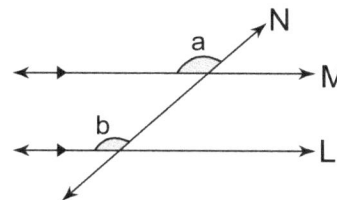

(19)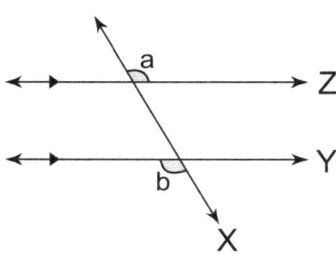

(20)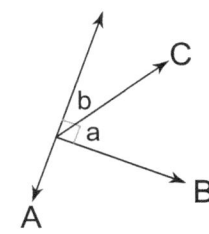

(21)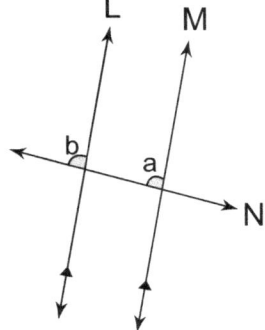

(22)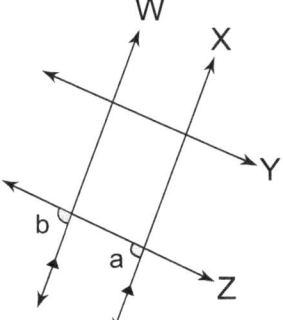

Identify the angles relationship from the below figures. Label them as complementary, supplementary, vertical, adjacent, alternate interior, corresponding, or alternate exterior angles.

(23)
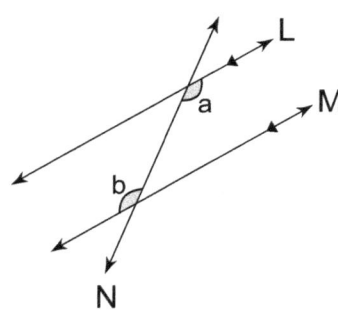

(24)
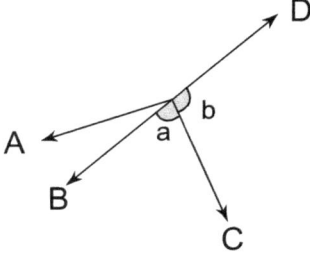

(25)
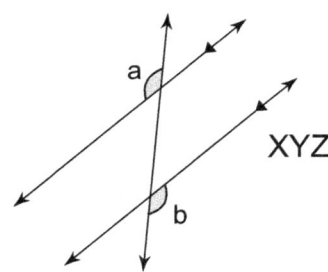
XYZ

(26)
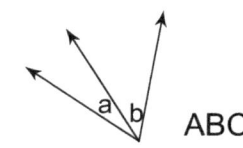
ABC

(27)
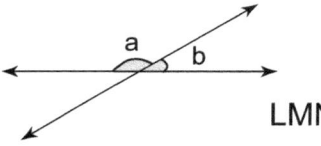
LMN

(28)
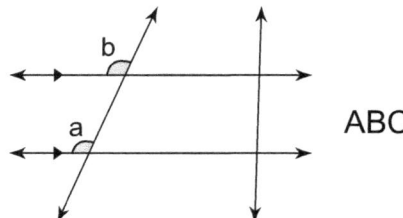
ABC

(29)
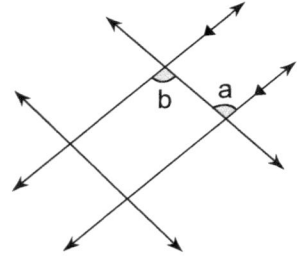

(30)
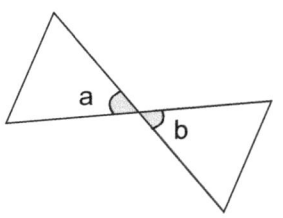

Find the missing value of b from the below figures.
Remember to use the angle relationships.

(31)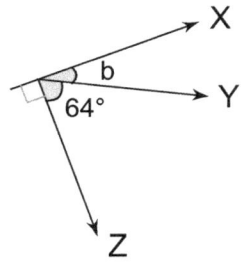

(32)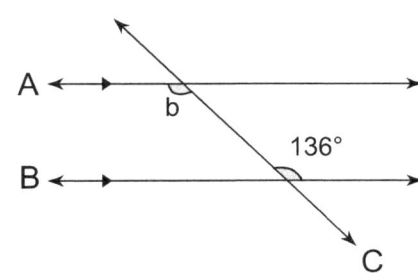

(33)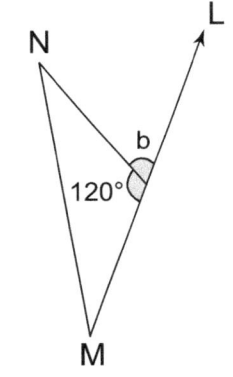

(34)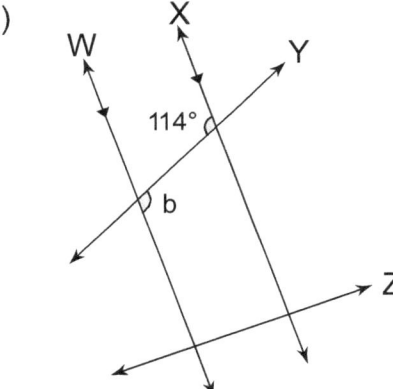

(35)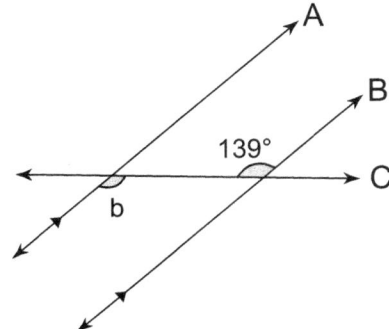

(36)

Find the missing value of b from the below figures.
Remember to use the angle relationships.

(37)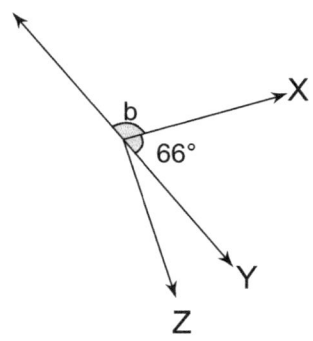

(38)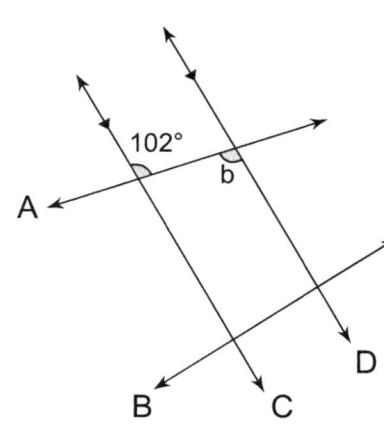

(39)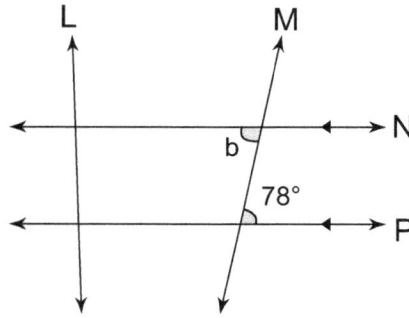

(40)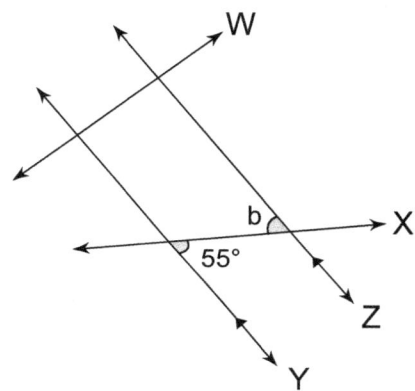

(41)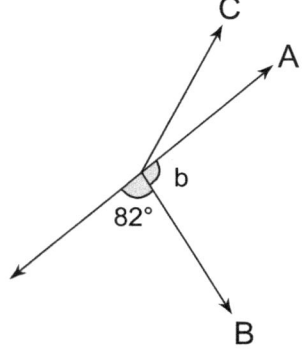

(42)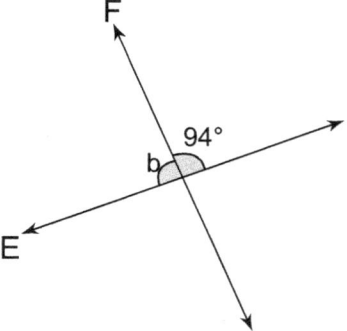

Find the missing value of b from the below figures.
Remember to use the angle relationships.

(43)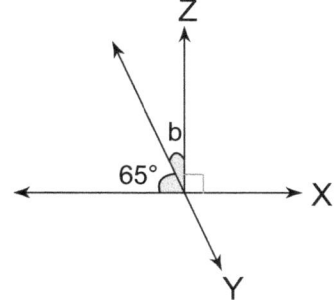

(44)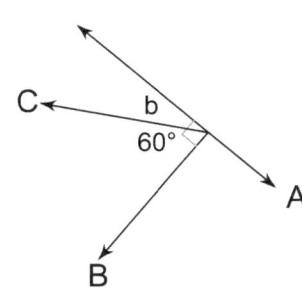

(45)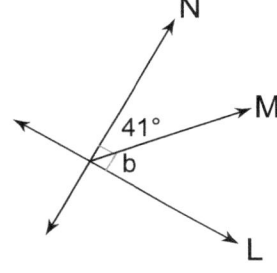

(46)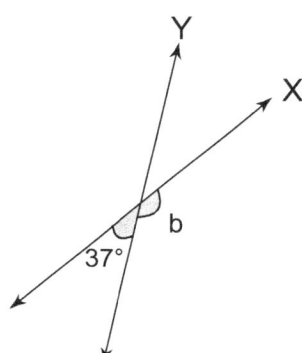

(47)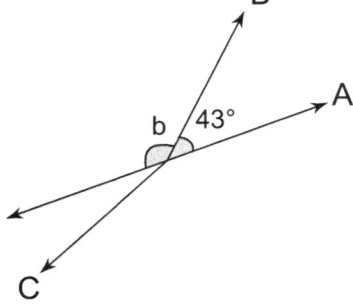

(48)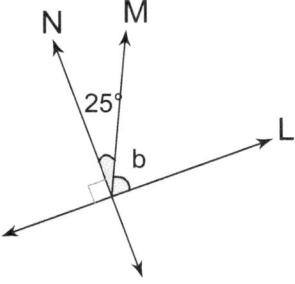

Find the missing value of **b** from the below figures.
Remember to use the angle relationships.

(49)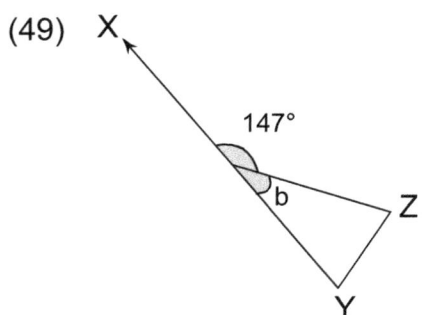

(50)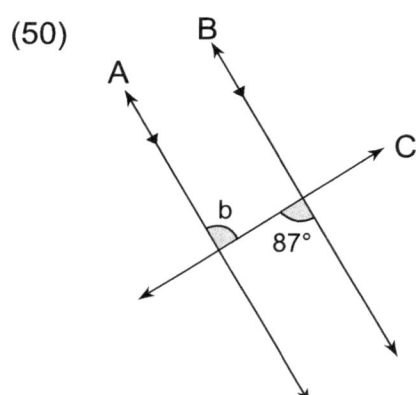

(51)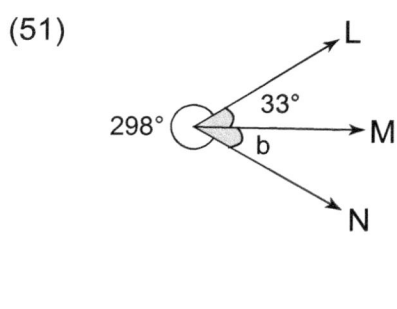

(52)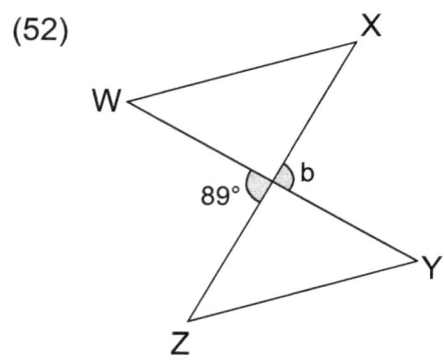

(53)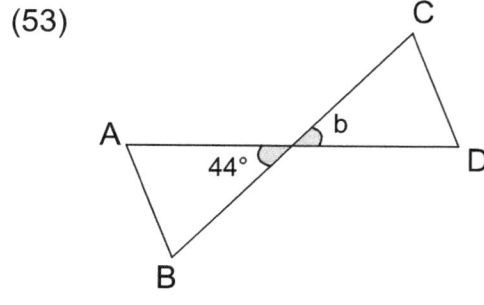

Find the missing value of b from the below figures.
Remember to use the angle relationships.

(54)

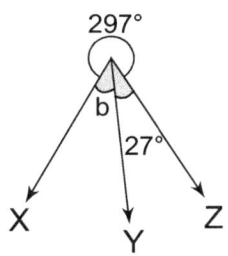

(55)

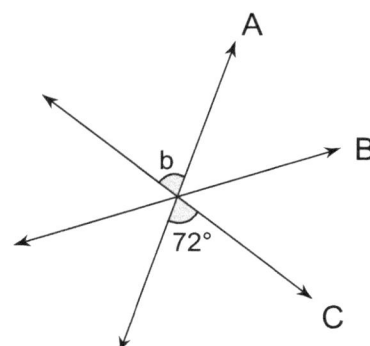

(56)

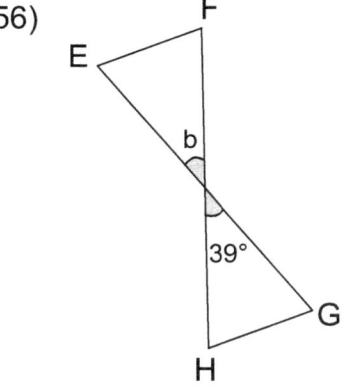

(57)

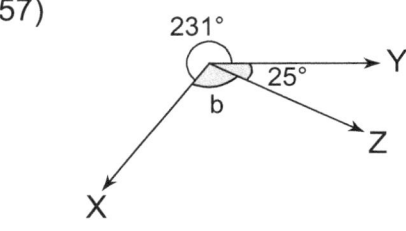

(58)

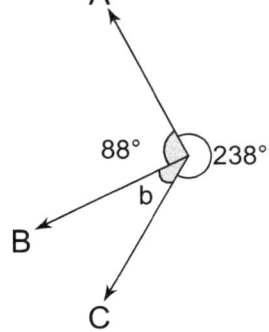

(59)

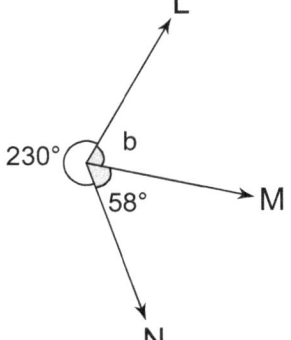

GEOMETRY

Basic Math

Find the missing value of **b** from the below figures.
Remember to use the angle relationships.

(60)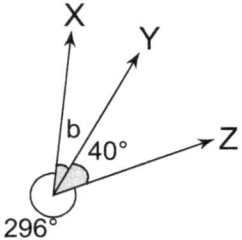

(61)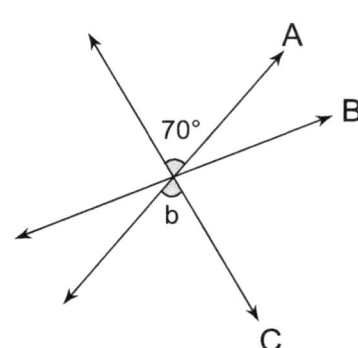

(62)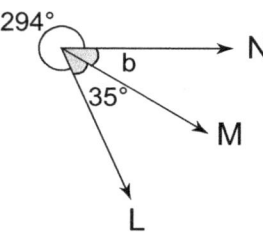

(63)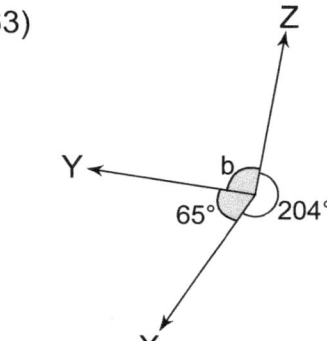

(64)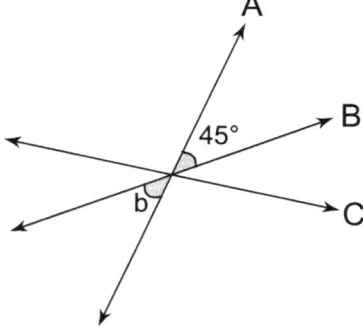

(65)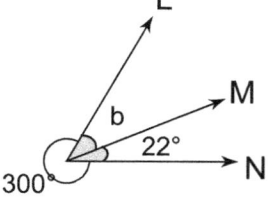

Find the missing value of *b* from the below figures.
Remember to use the angle relationships.

(66)
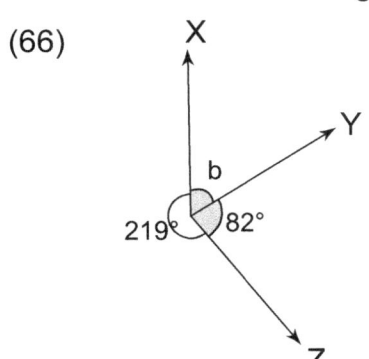

(67)
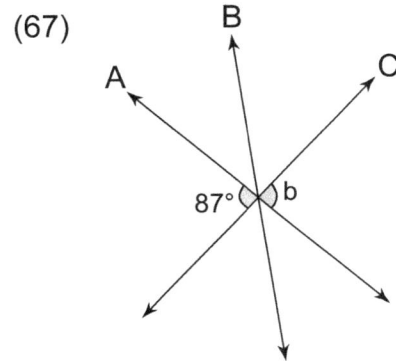

(68)
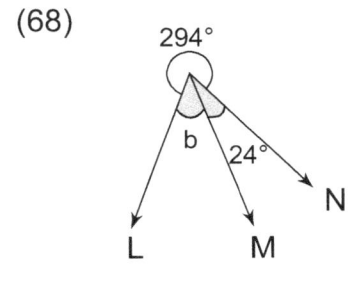

(69)
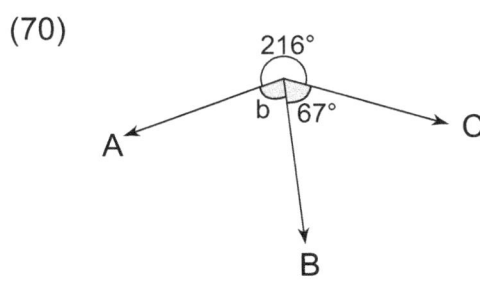

(70)

GEOMETRY

Basic Math

Find the missing value of **b** from the below figures.
Remember to use the angle relationships.

(71)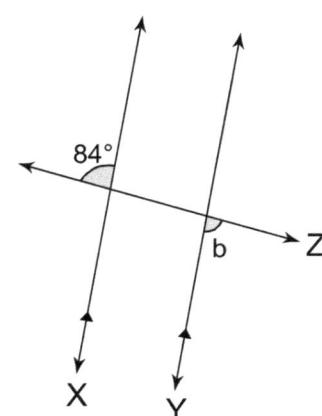

(72)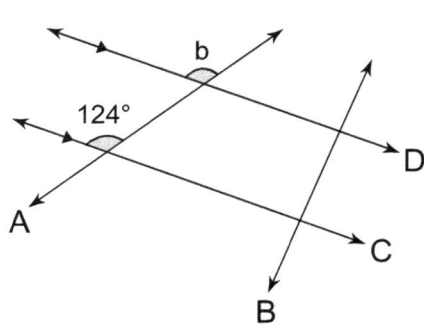

(73)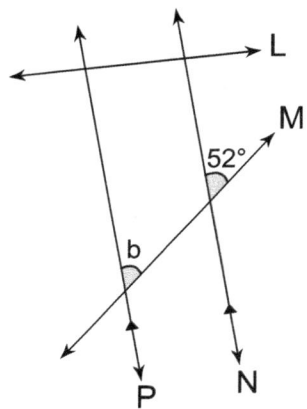

(74)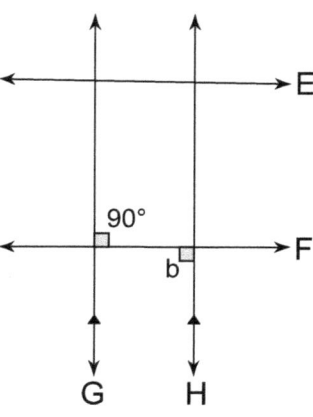

(75)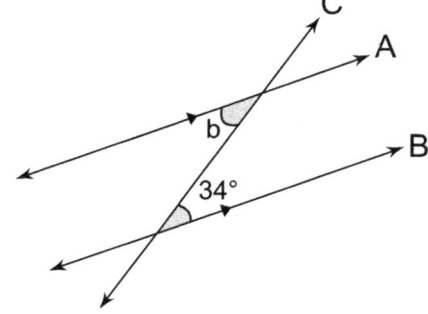

(76)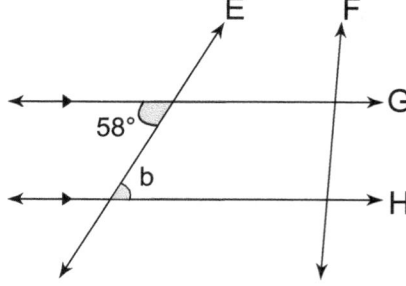

GEOMETRY

Basic Math

Find the missing value of b from the below figures.
Remember to use the angle relationships.

(77)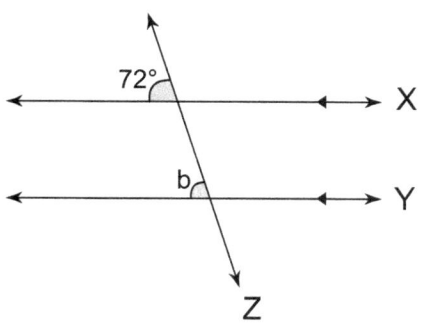

(78)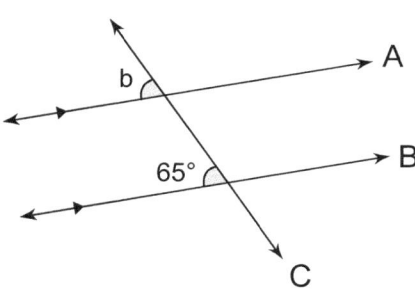

(79)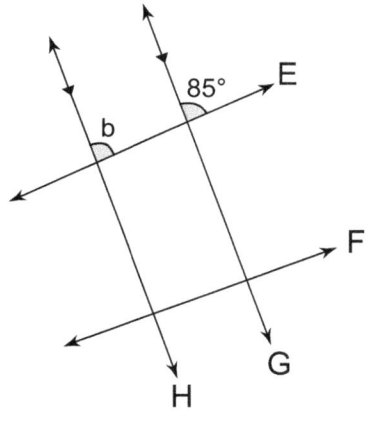

(80)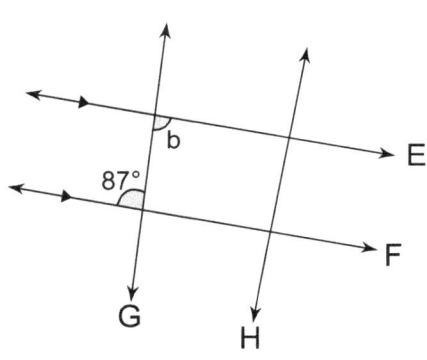

(81)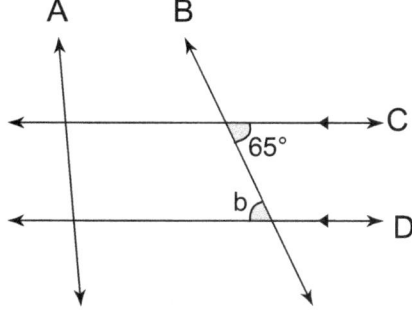

(82)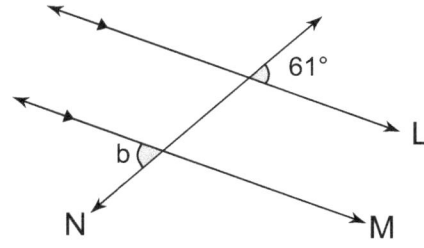

©All rights reserved-Math-Knots LLC., VA-USA
For more visit www.a4ace.com

45

www.math-knots.com

Find the missing value of **b** from the below figures.
Remember to use the angle relationships.

(83)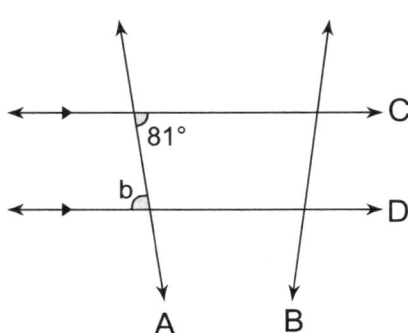

(84)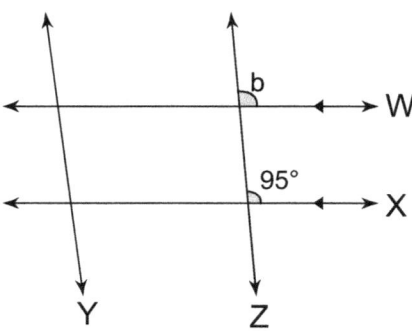

(85)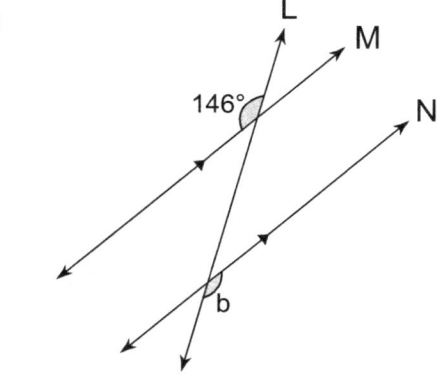

(86)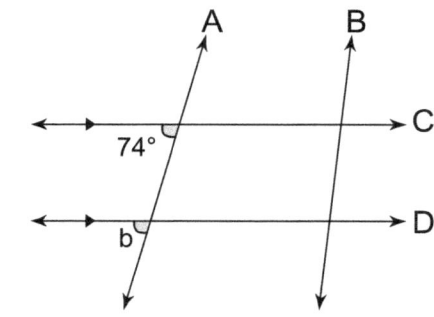

(87)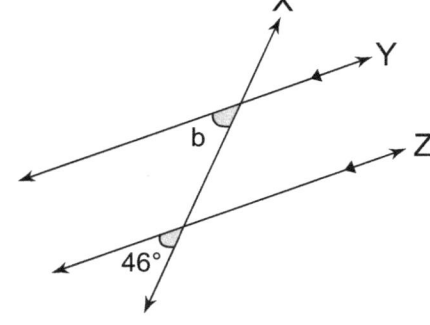

(88)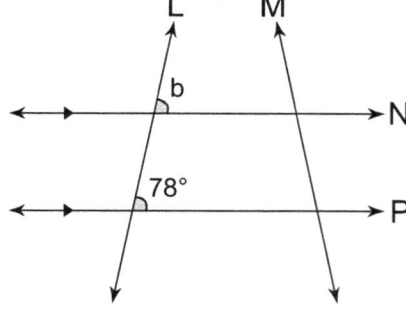

Find the missing value of b from the below figures.
Remember to use the angle relationships.

(89)
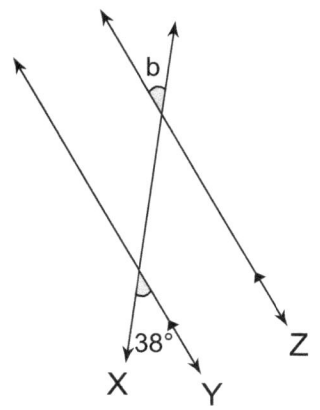

(90)
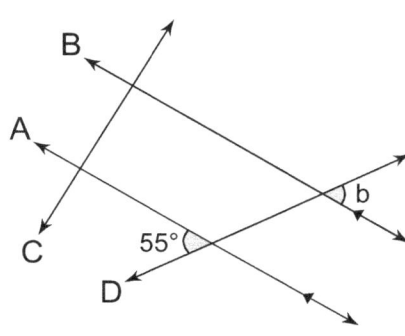

(91)
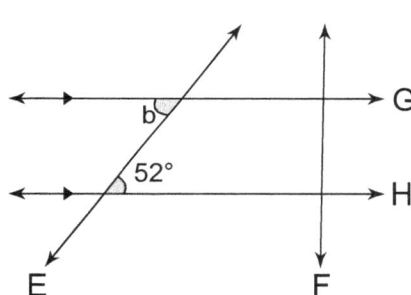

(92)
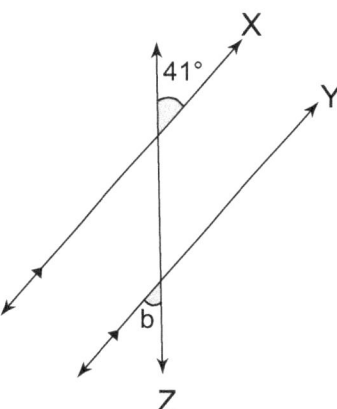

(93)
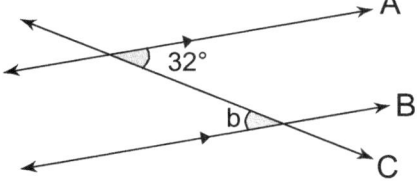

(94)
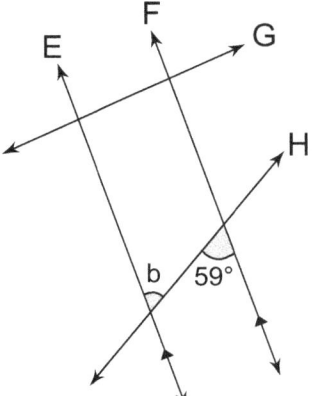

Find the missing value of ∠b from the below figures.
Remember to use the angle relationships.

(95)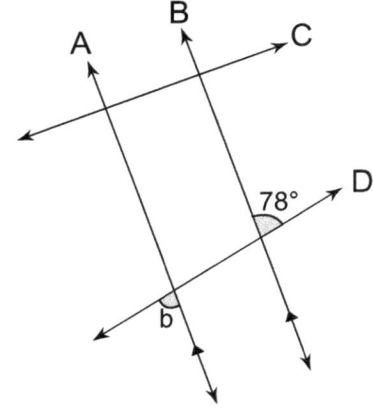

(96)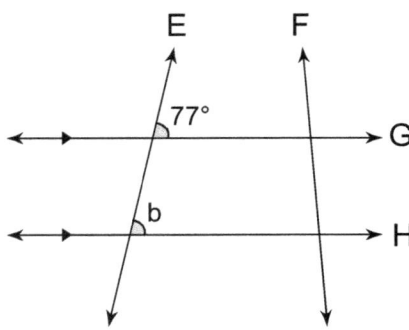

(97)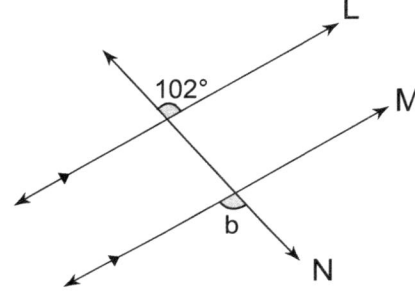

(98)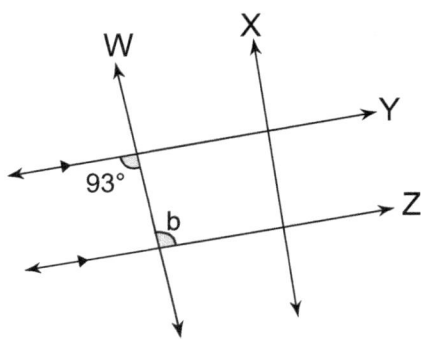

(99)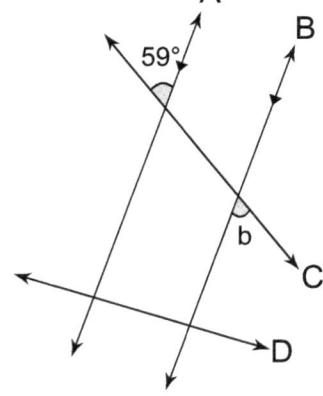

(100)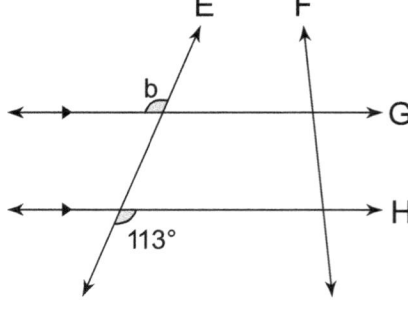

Find the missing value of ∠x from the below figures.
Remember to use the angle relationships.

(101)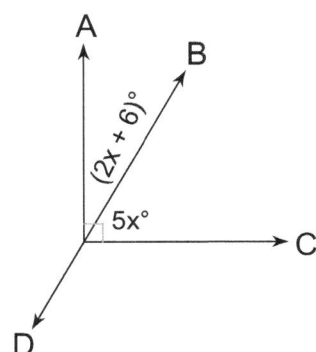

(102)

(103)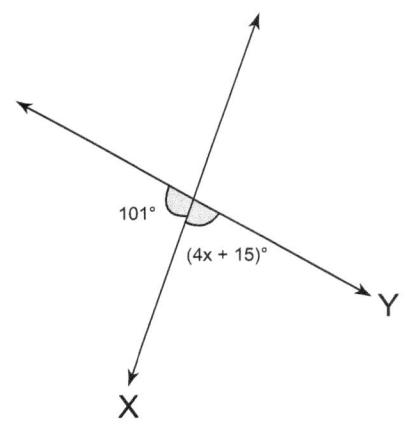

(104)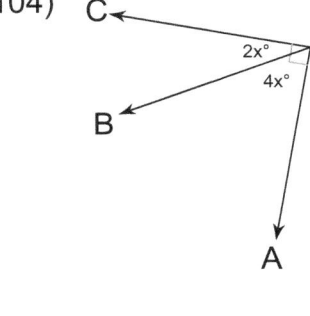

(105)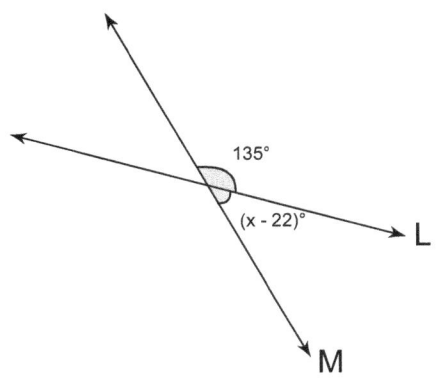

(106)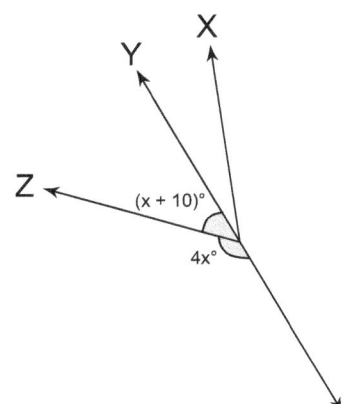

Find the missing value of x from the below figures.
Remember to use the angle relationships.

(107)
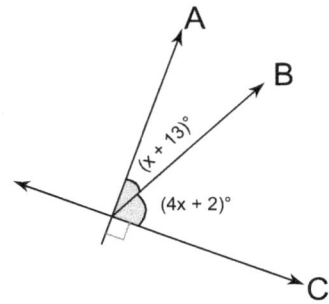

(108)
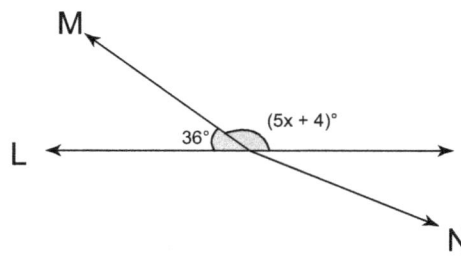

(109)
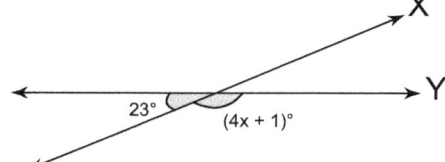

(110)
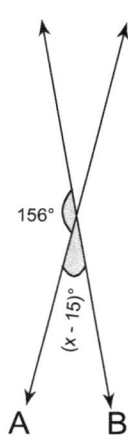

(111)
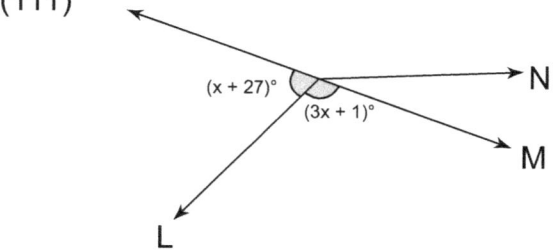

(112)
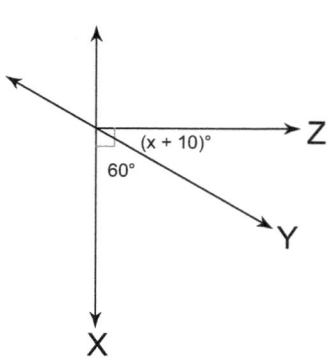

Find the missing value of x from the below figures.
Remember to use the angle relationships.

(113)

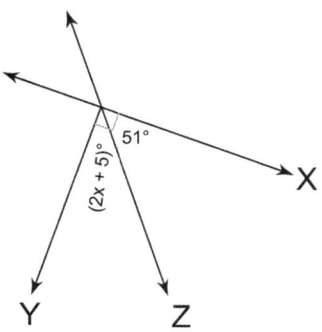

(114)

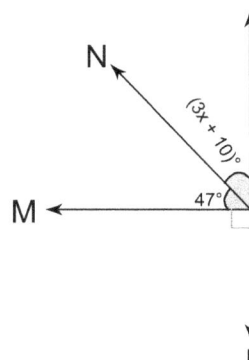

(115)

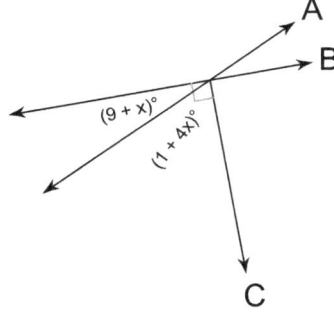

(116)

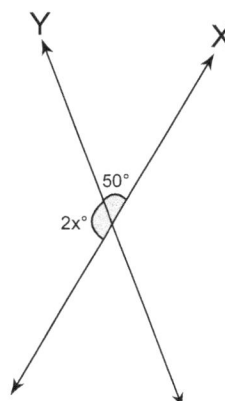

(117)

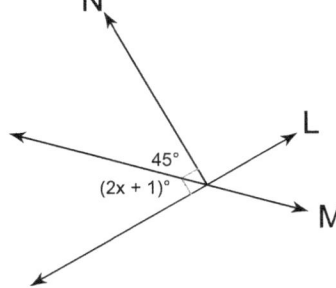

(118)

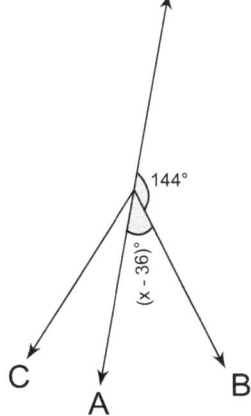

 **GEOMETRY**

Basic Math

Find the missing value of x from the below figures.
Remember to use the angle relationships.

(119)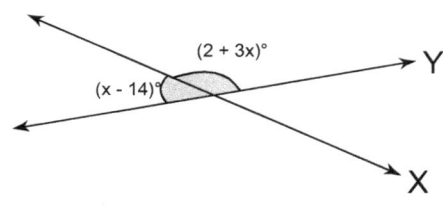

(120)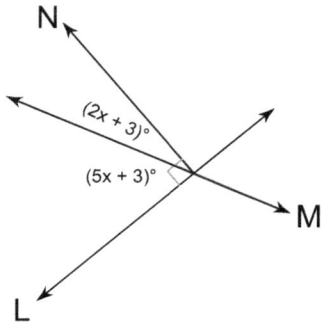

(121)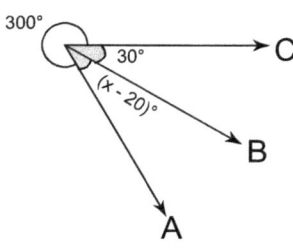

(122)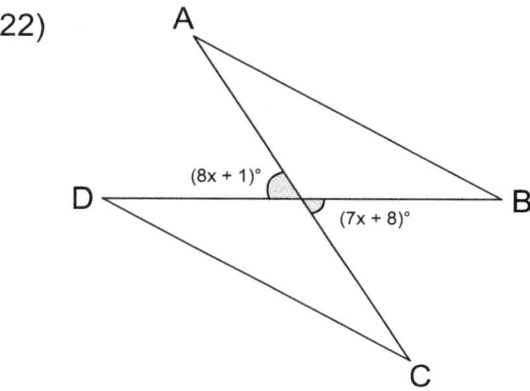

(123)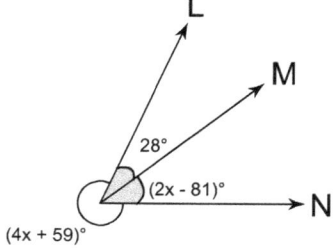

(124)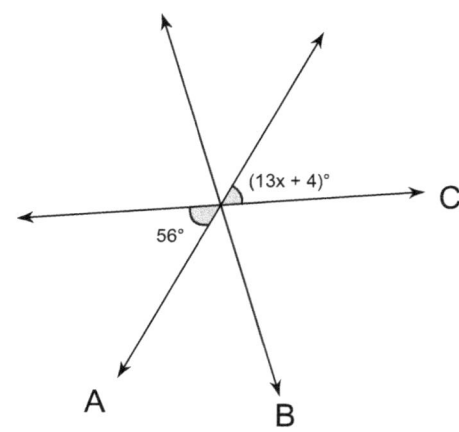

Find the missing value of |x from the below figures.
Remember to use the angle relationships.

(125)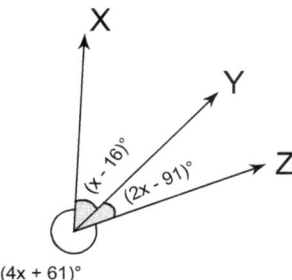

(126)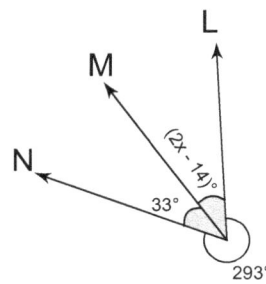

(127)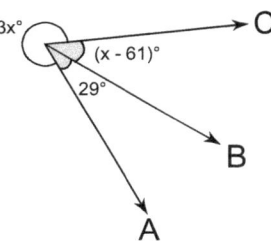

(128)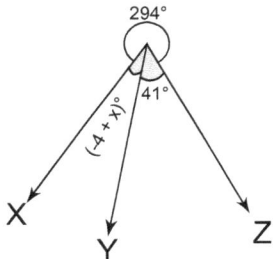

(129)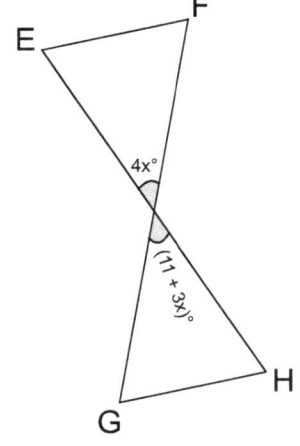

(130)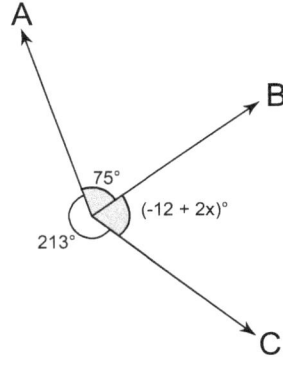

Find the missing value of |x from the below figures.
Remember to use the angle relationships.

(131)

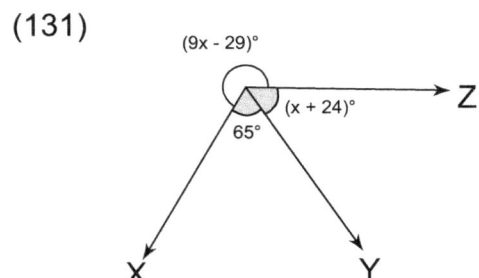

(132)

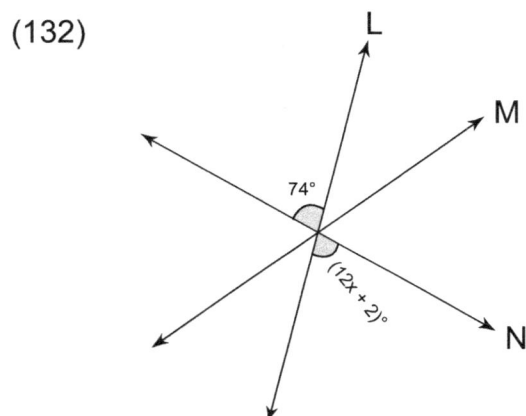

(133)

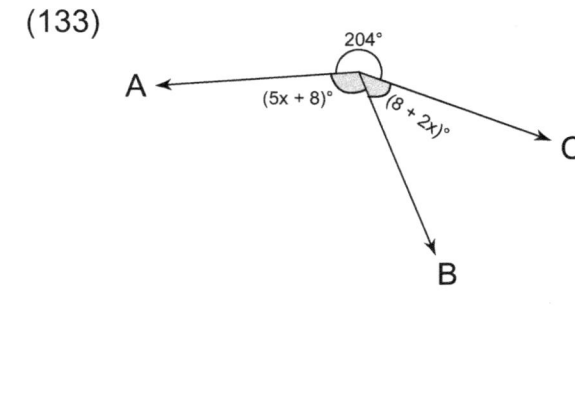

(134)

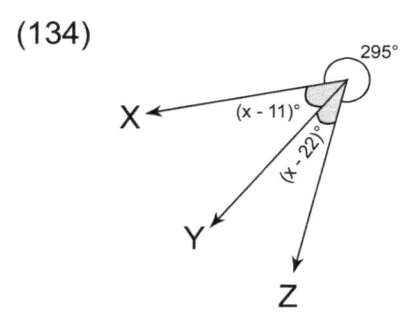

(135)

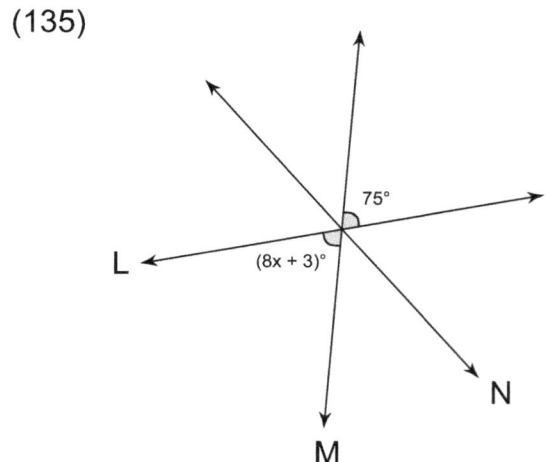

GEOMETRY

Basic Math

Find the missing value of <u>x</u> from the below figures.
Remember to use the angle relationships.

(136)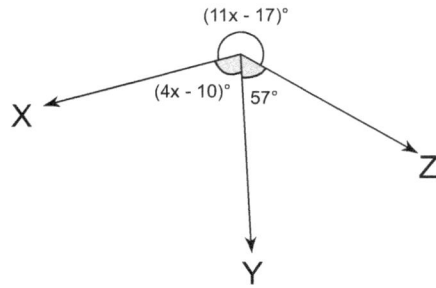

(137)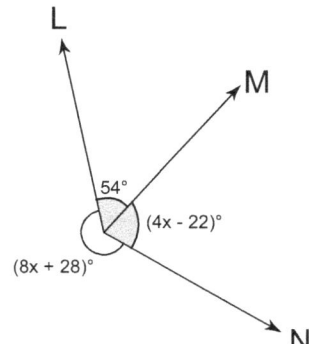

(138)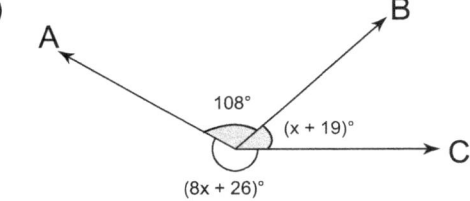

(139)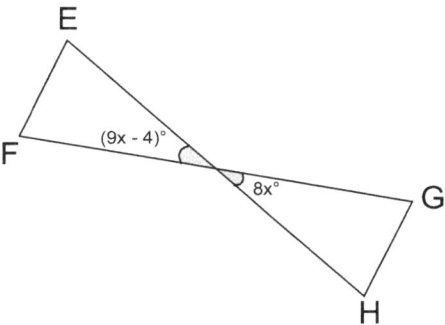

(140)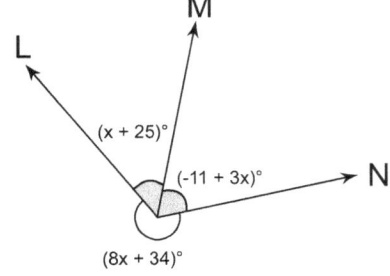

(141)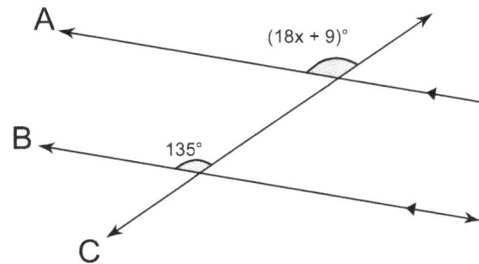

©All rights reserved-Math-Knots LLC., VA-USA
For more visit www.a4ace.com

www.math-knots.com

GEOMETRY

Find the missing value of x from the below figures.
Remember to use the angle relationships.

(142)
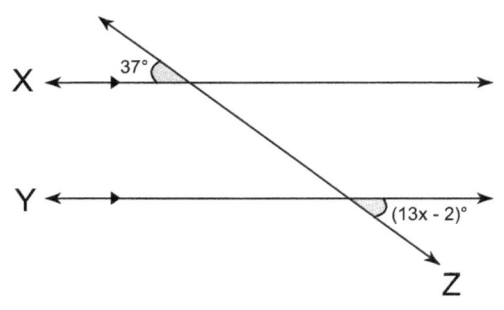

(143)
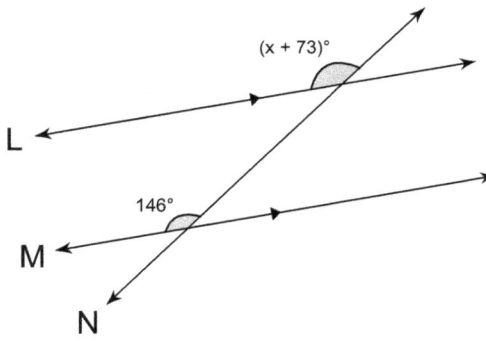

(144)
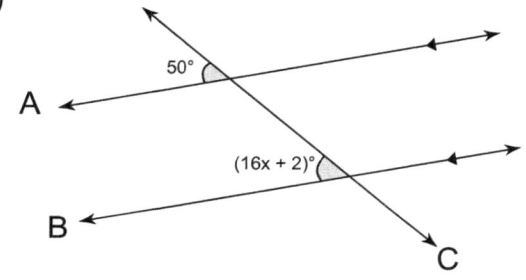

(145)
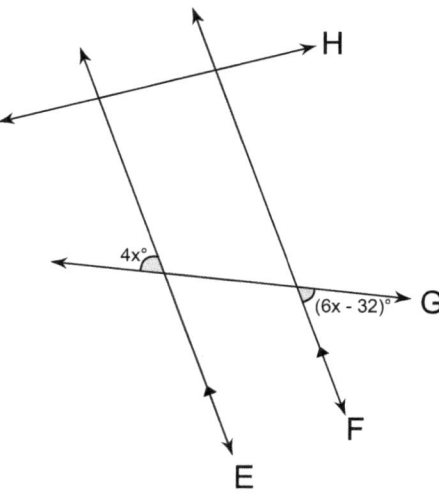

(146)
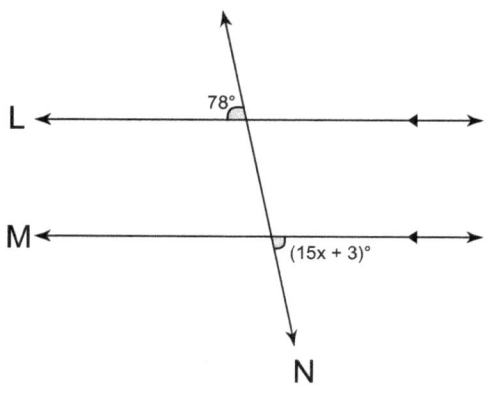

(147)
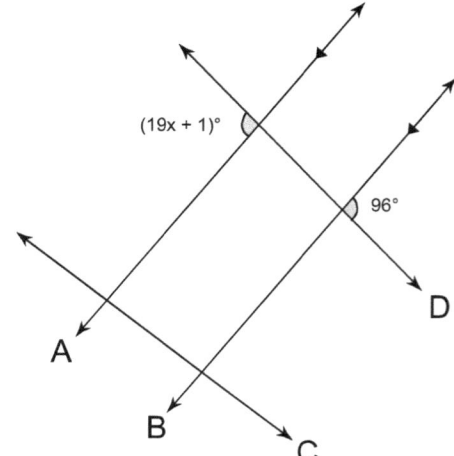

Find the missing value of |x from the below figures. Remember to use the angle relationships.

(148)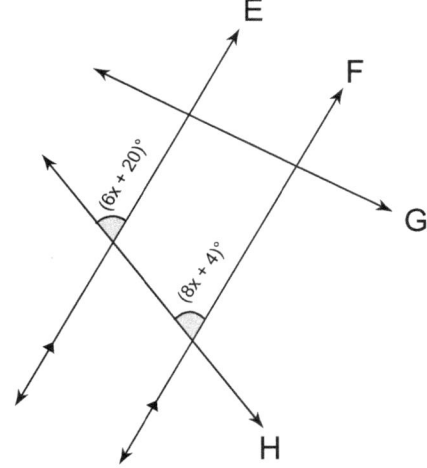

(149)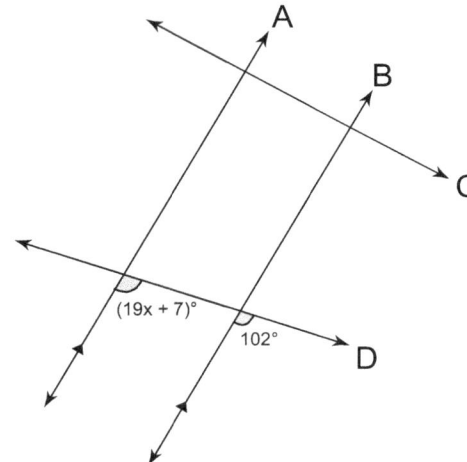

(150)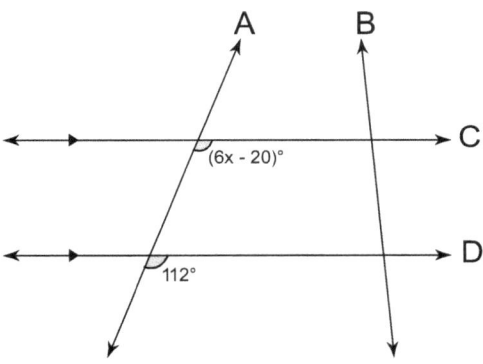

(151)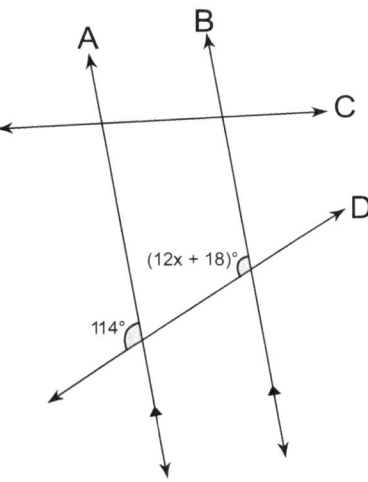

(152)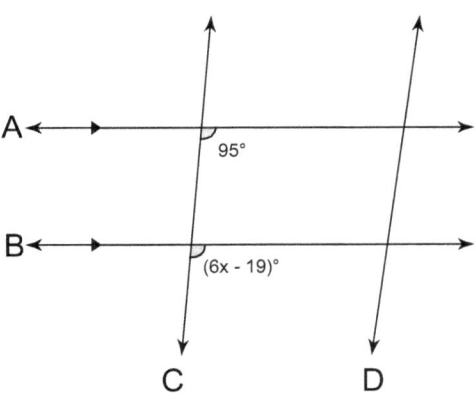

(153)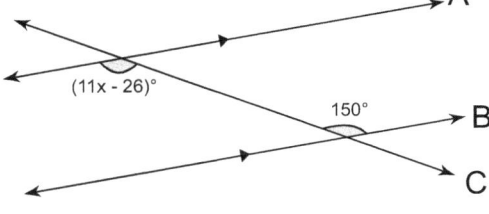

Find the missing value of x from the below figures.
Remember to use the angle relationships.

(154)

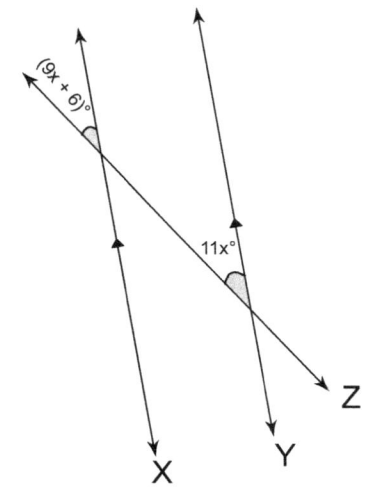

(155)

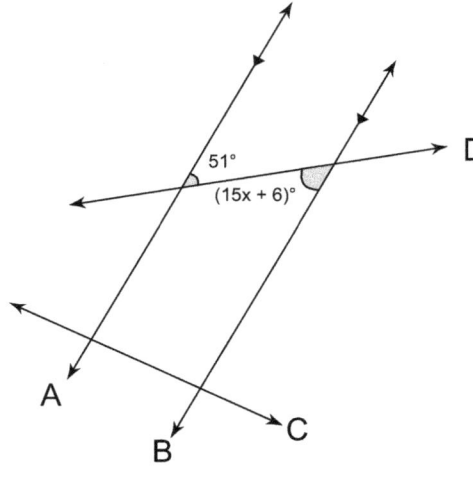

(156)

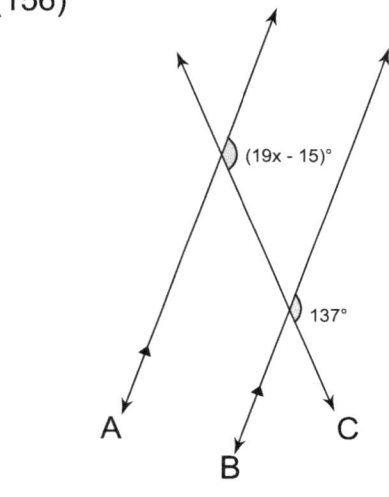

(157)

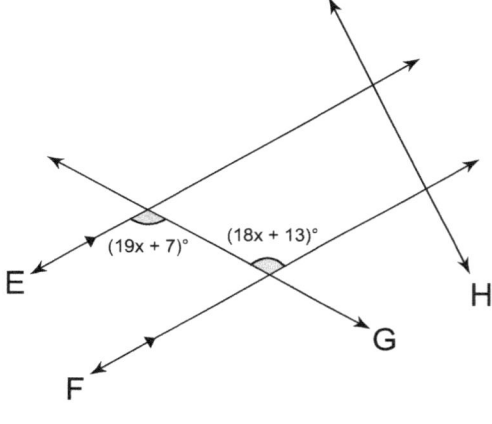

(158)

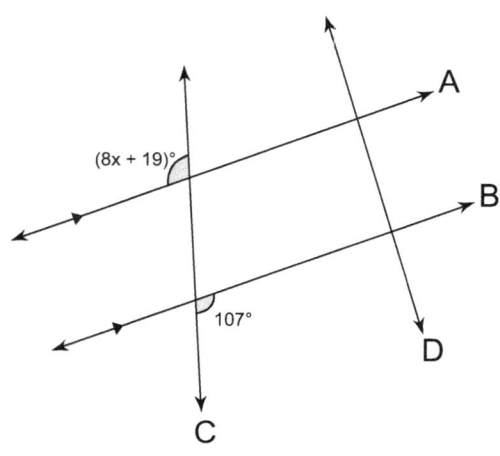

(159)

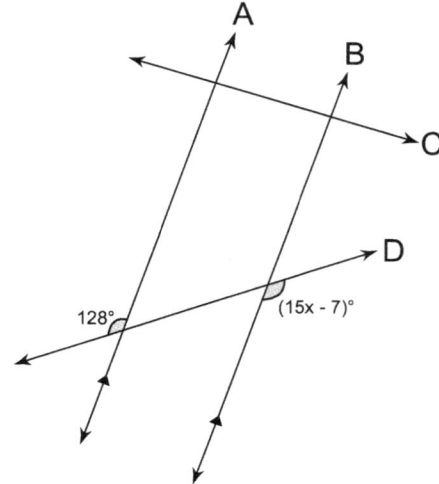

GEOMETRY

Basic Math

Find the missing value of x from the below figures.
Remember to use the angle relationships.

(160)

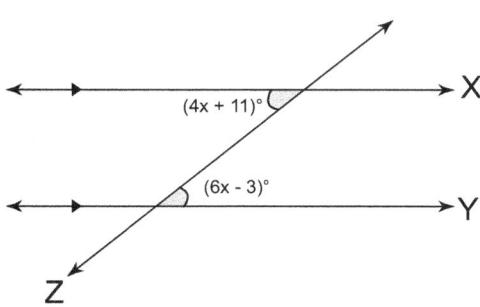

(161)

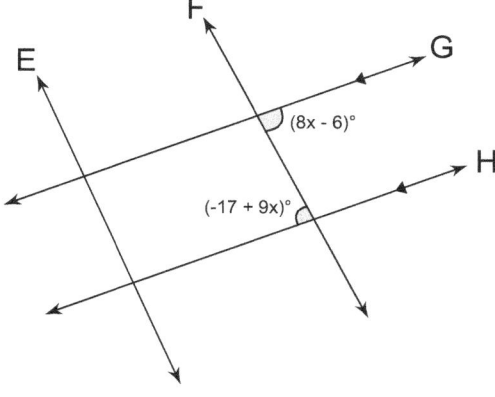

(162)

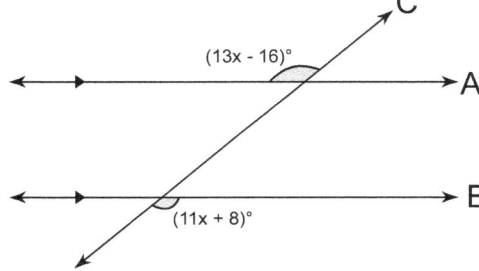

(163)

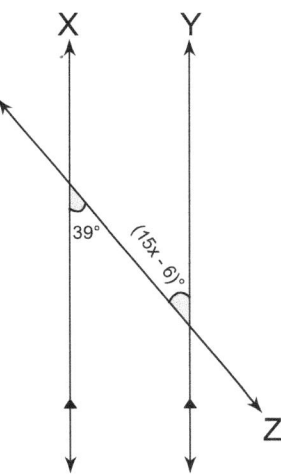

(164)

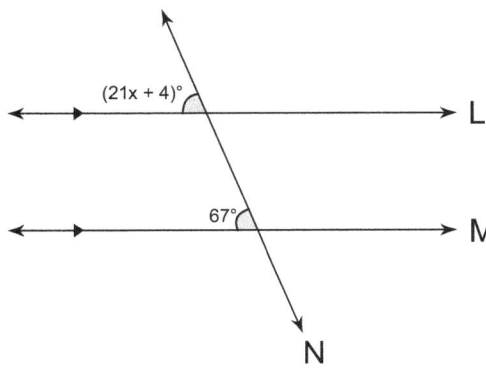

(165)

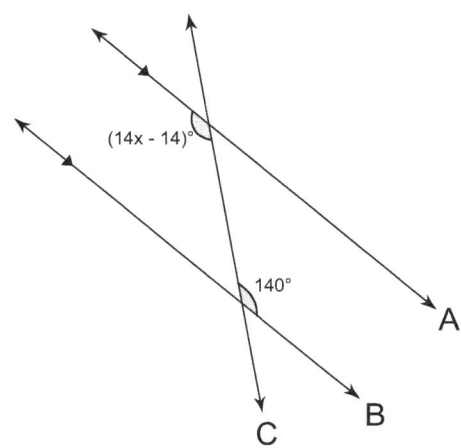

©All rights reserved-Math-Knots LLC., VA-USA
For more visit www.a4ace.com
59
www.math-knots.com

Find the missing value of x from the below figures.
Remember to use the angle relationships.

(166)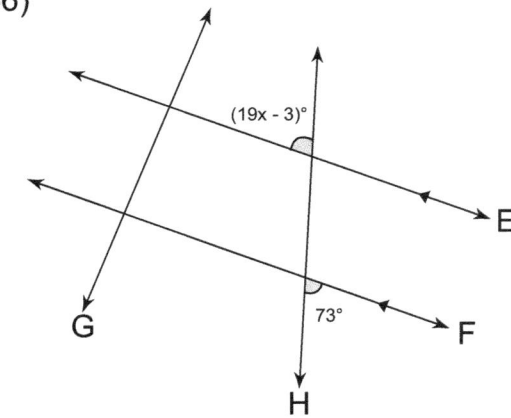

(167)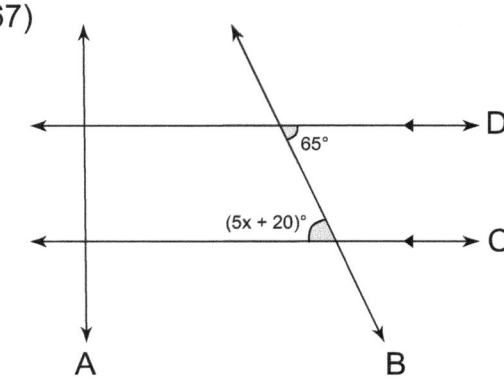

(168)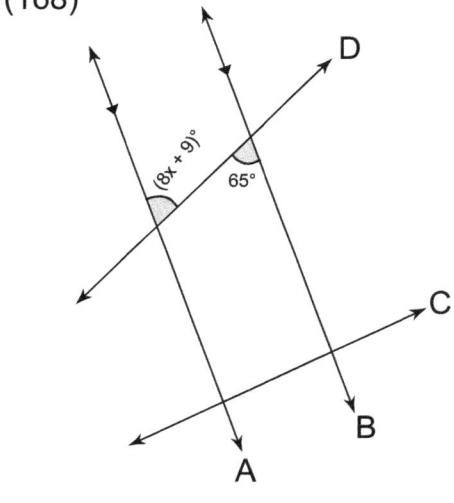

(169)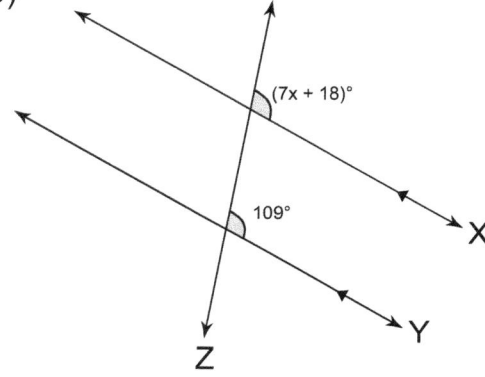

(170)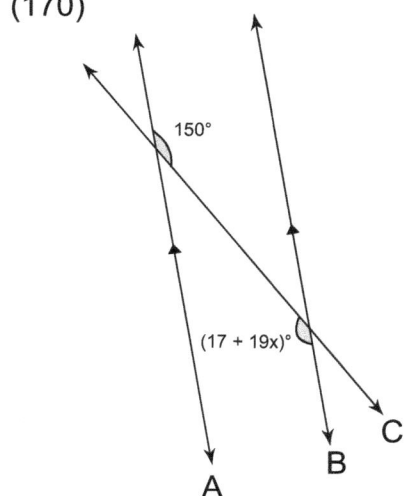

GEOMETRY

Basic Math

Find the missing value of *b* from the below figures.
Remember to use the angle relationships.

(171)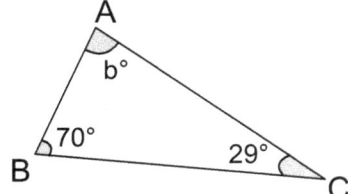

(172)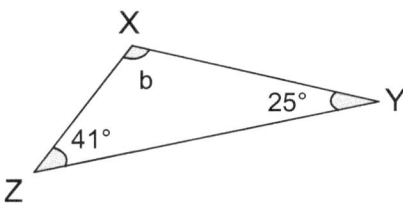

(173)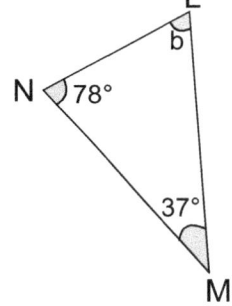

(174)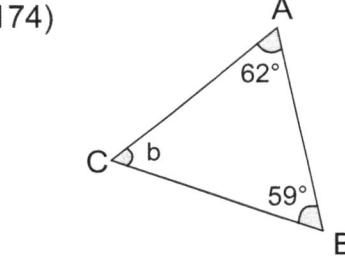

(175)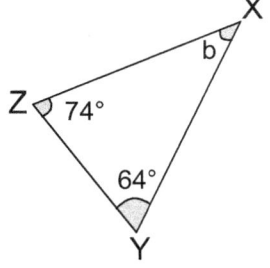

(176)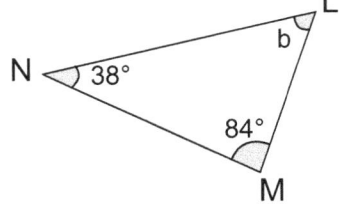

Find the missing value of **b** from the below figures.
Remember to use the angle relationships.

(177)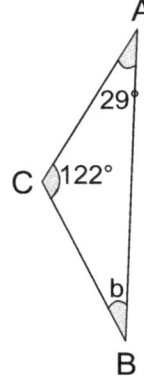

(178)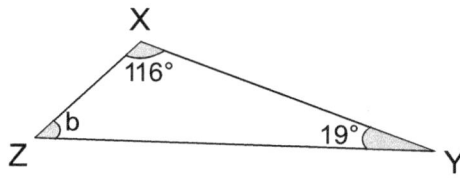

(179)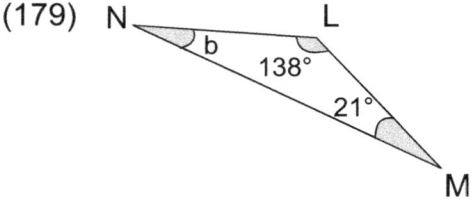

(180)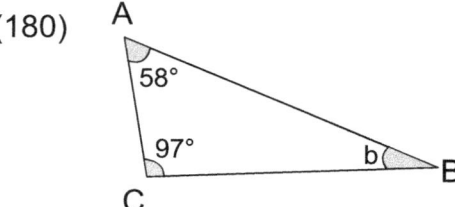

(181)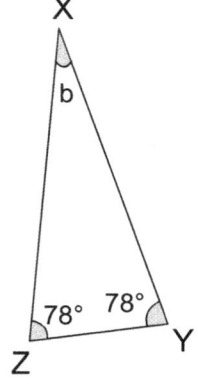

(182)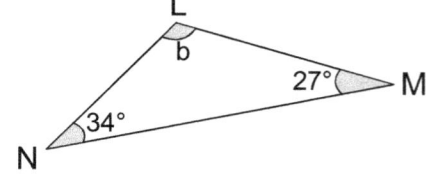

 **GEOMETRY**

Basic Math

Find the missing value of b from the below figures.
Remember to use the angle relationships.

(183)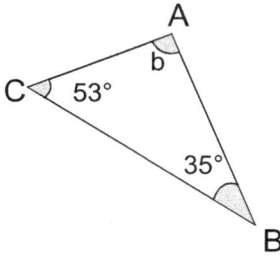

(184)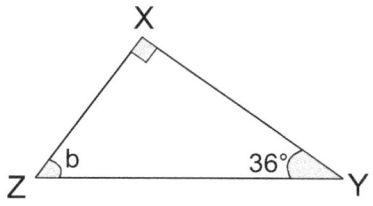

(185)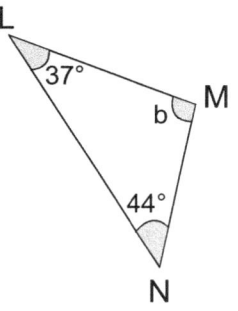

(186)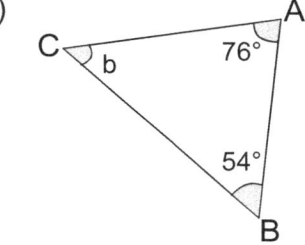

(187)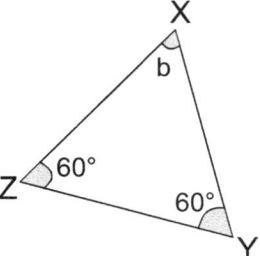

(188)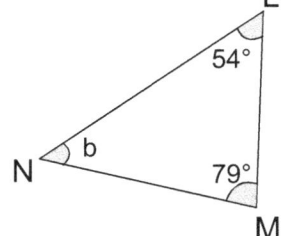

Find the missing value of b from the below figures.
Remember to use the angle relationships.

(189)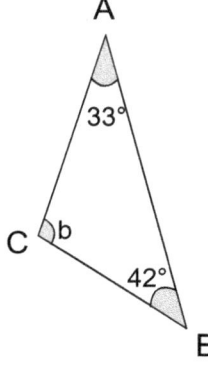

(190)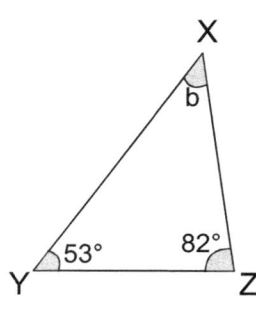

(191)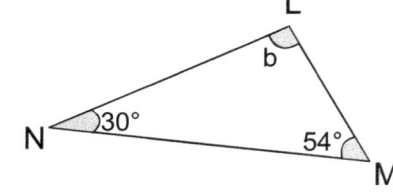

(192)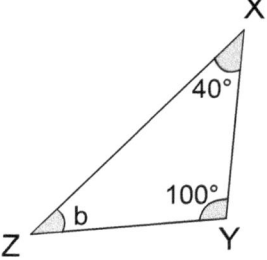

(193)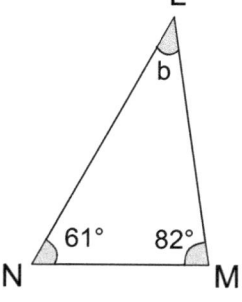

(194)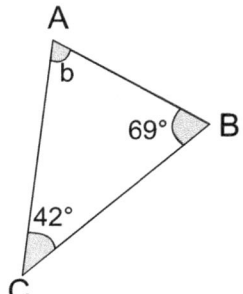

Find the missing value of b from the below figures.
Remember to use the angle relationships.

(195)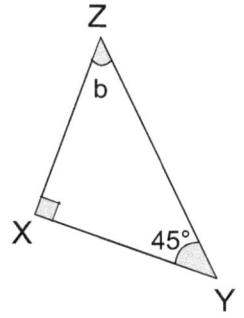

(196)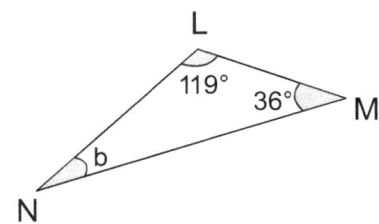

(197)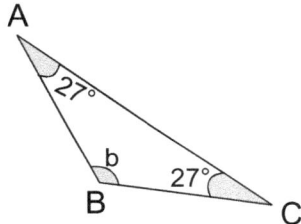

(198)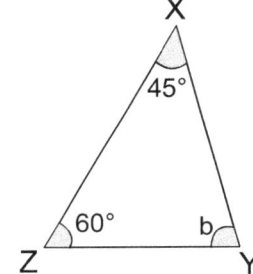

(199)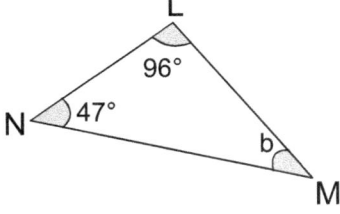

(200)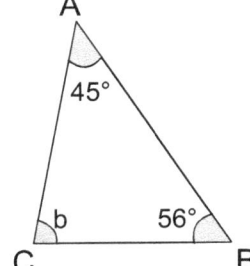

Find the missing value of x from the below figures.
Remember to use the angle relationships.

(201)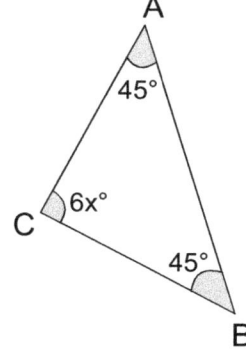

(202)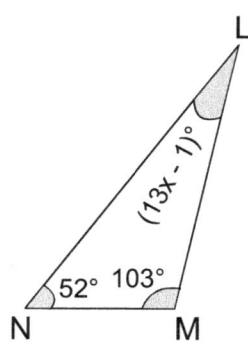

(203)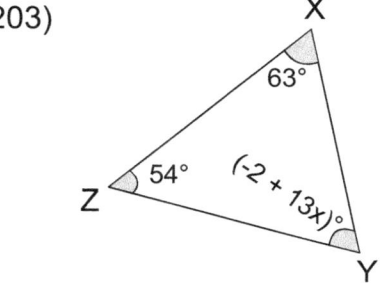

(204)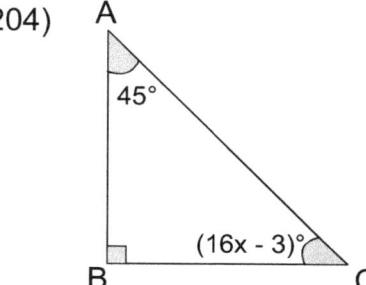

(205)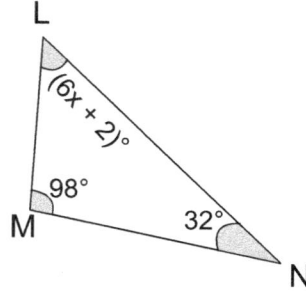

(206)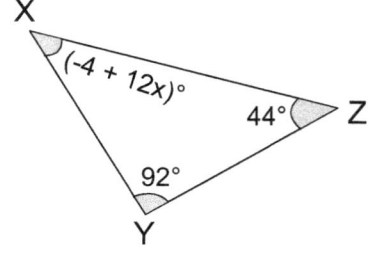

Find the missing value of x from the below figures.
Remember to use the angle relationships.

(207)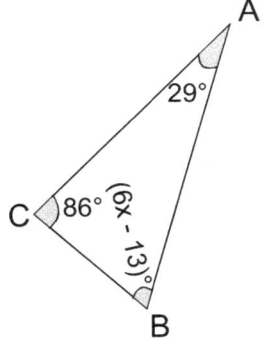

(208)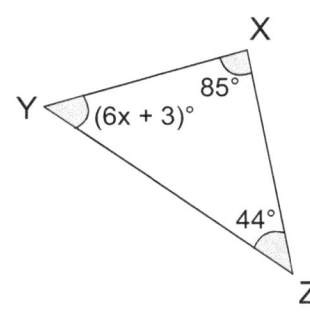

(209)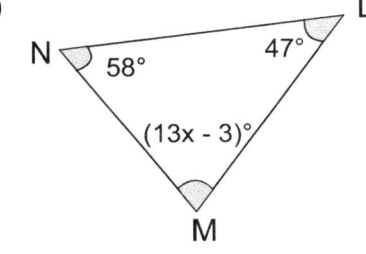

(210)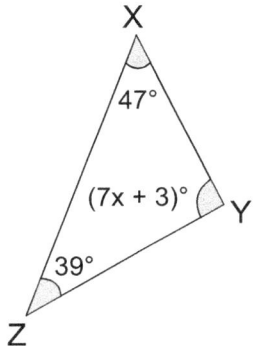

(211)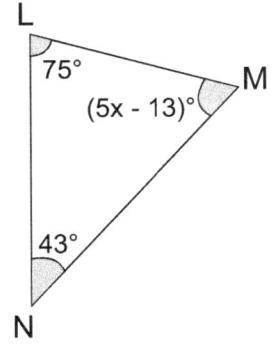

(212)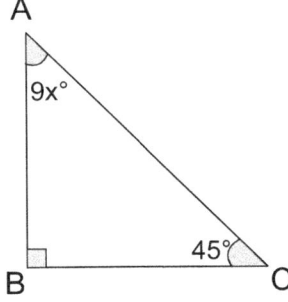

GEOMETRY

Find the missing value of x from the below figures.
Remember to use the angle relationships.

(213)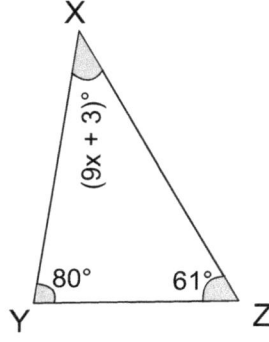

(214)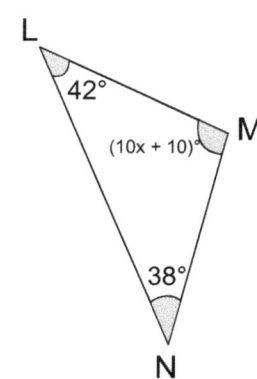

(215)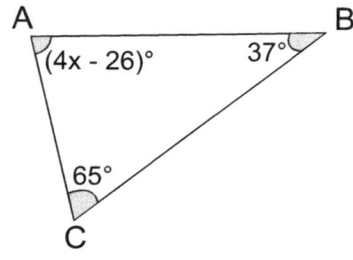

(216)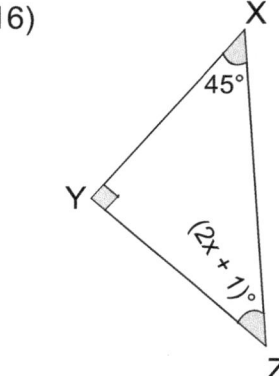

(217)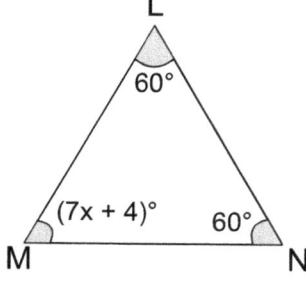

(218)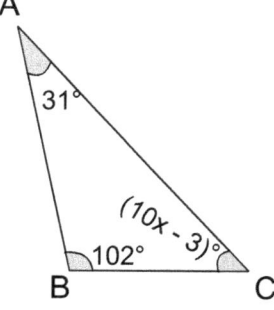

Find the missing value of |x from the below figures.
Remember to use the angle relationships.

(219)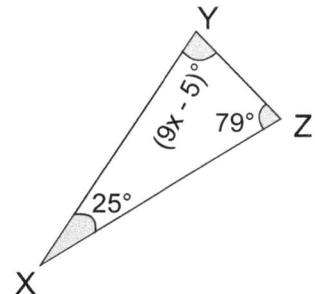

(220)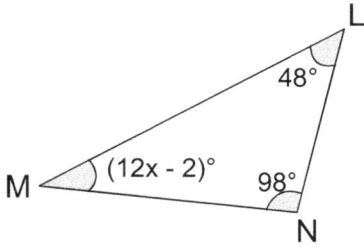

(221)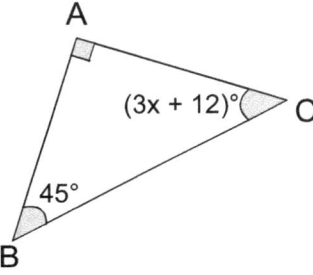

(222)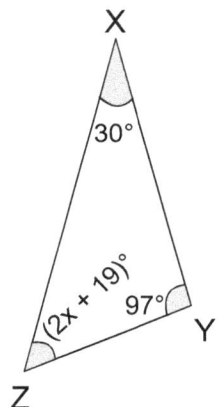

(223)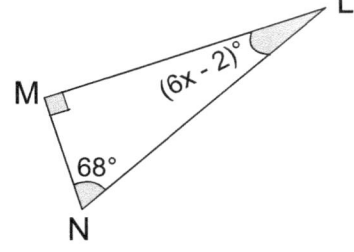

(224)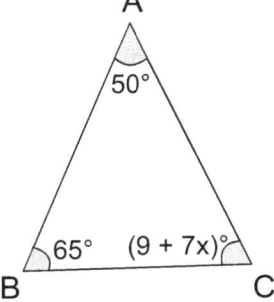

GEOMETRY

Basic Math

Find the missing value of x from the below figures.
Remember to use the angle relationships.

(225)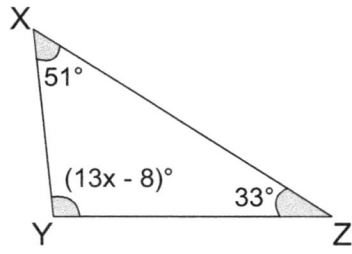

(226)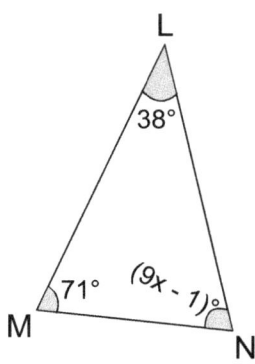

(227)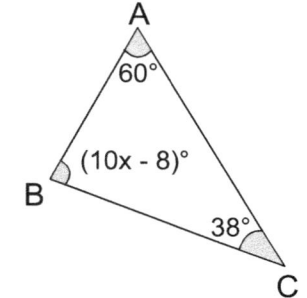

(228)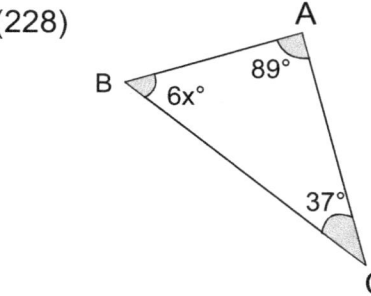

(229)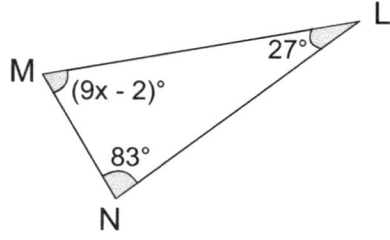

(230)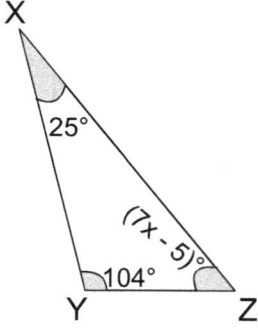

Find the missing value of x from the below figures.
Remember to use the angle relationships.

(231)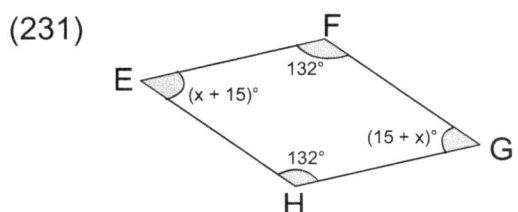

(232)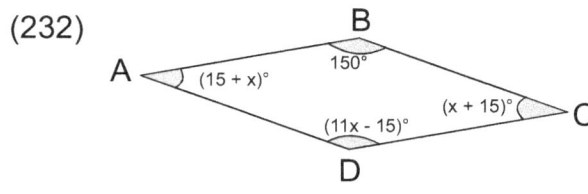

(233)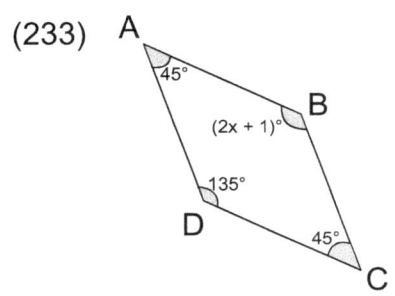

(234)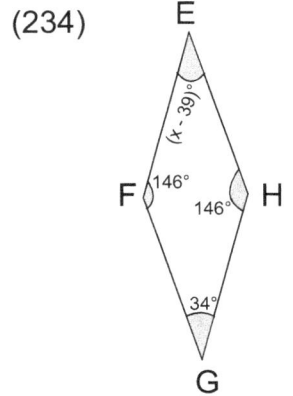

(235)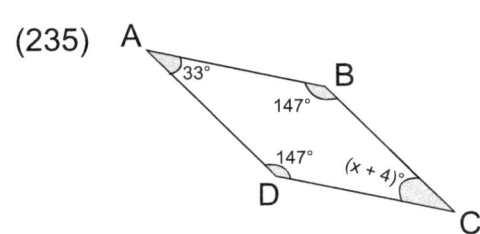

(236)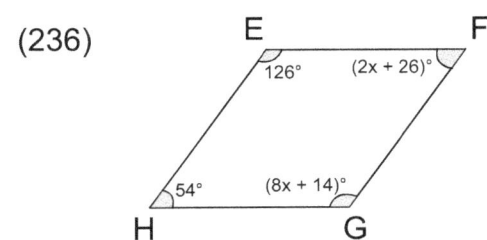

Find the missing value of |x from the below figures.
Remember to use the angle relationships.

(237) Quadrilateral ABCD with angle A = (6 + 7x)°, angle B = 90°, angle C = (6x + 18)°, angle D = (7x + 6)°.

(238) Quadrilateral EFGH with angle E = (12x − 6)°, angle F = 90°, angle G = (12x − 6)°, angle H = 90°.

(239) Triangle-like figure HIJK with angle H = (x + 3)°, angle I = 142°, angle K = 142°, angle J = 38°.

(240) Parallelogram PQRS with angle P = 130°, angle Q = 50°, angle R = 130°, angle S = (2x − 14)°.

(241) Parallelogram EFGH with angle E = (7x − 18)°, angle F = 44°, angle G = (7x − 18)°, angle H = 44°.

(242) Parallelogram ABCD with angle A = 11x°, angle B = (11 + 2x)°, angle C = (12x − 13)°, angle D = 37°.

Basic Math

Find the missing value of x from the below figures.
Remember to use the angle relationships.

(243)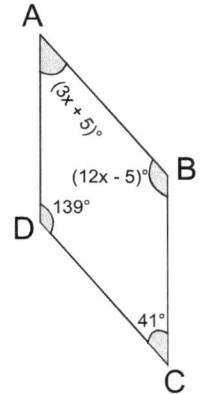

A, (3x + 5)°
B, (12x − 5)°
D, 139°
C, 41°

(244)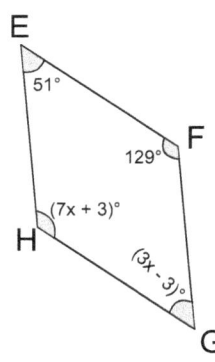

E, 51°
F, 129°
H, (7x + 3)°
G, (3x − 3)°

(245)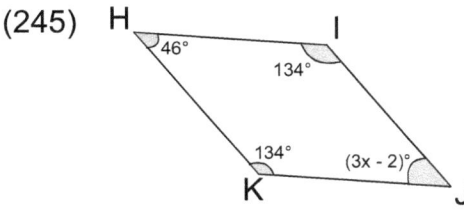

H, 46°
I, 134°
K, 134°
J, (3x − 2)°

(246)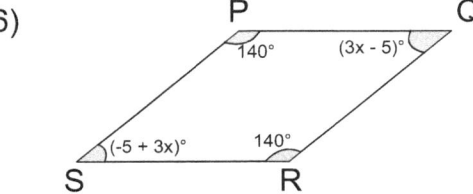

P, 140°
Q, (3x − 5)°
S, (−5 + 3x)°
R, 140°

(247)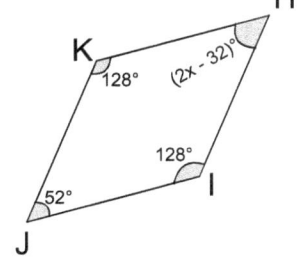

K, 128°
H, (2x − 32)°
I, 128°
J, 52°

(248)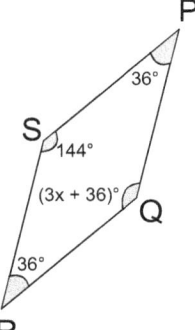

P, 36°
S, 144°
Q, (3x + 36)°
R, 36°

GEOMETRY

Find the missing value of x from the below figures.
Remember to use the angle relationships.

(249)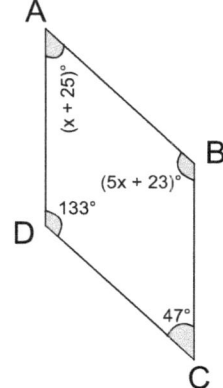

(250)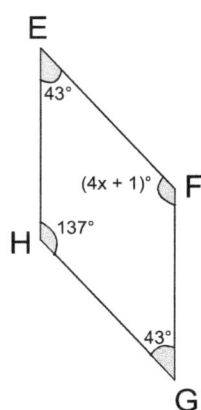

(251)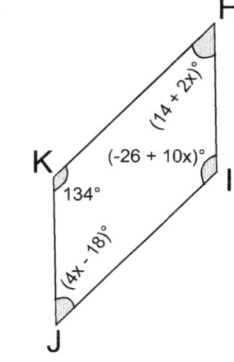

(252)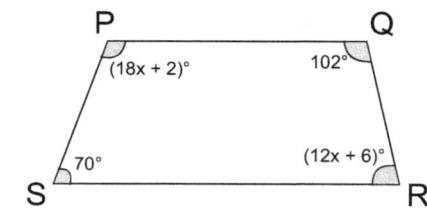

(253)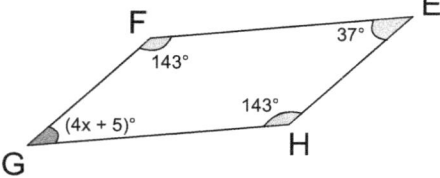

(254)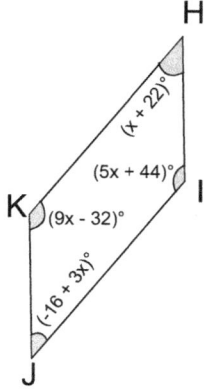

GEOMETRY

Basic Math

Find the missing value of |x from the below figures.
Remember to use the angle relationships.

(255)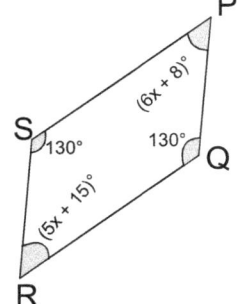

(256)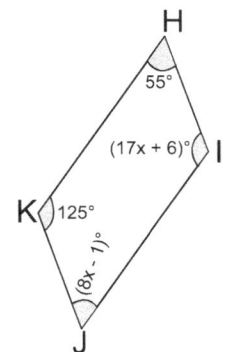

(257)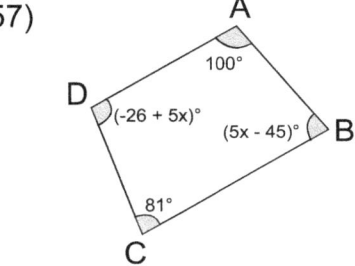

(258)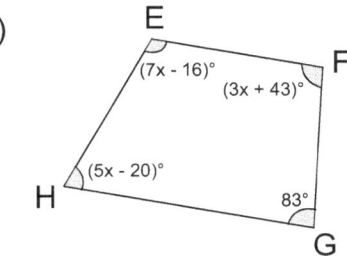

(259)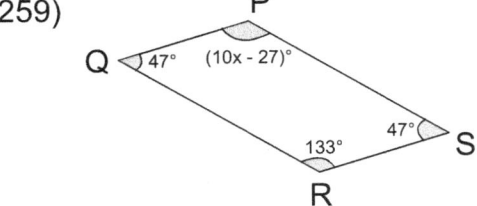

(260)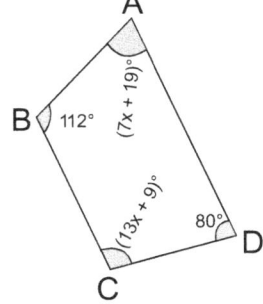

Find the missing value of |x from the below figures.
Remember to use the angle relationships.

(261)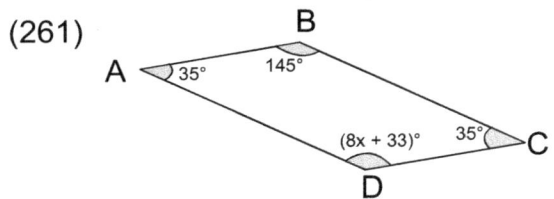

(262)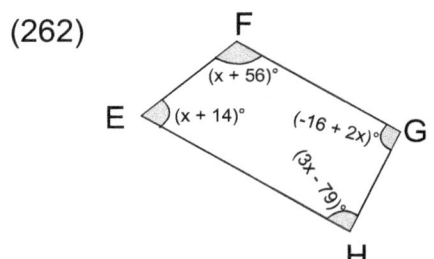

(263)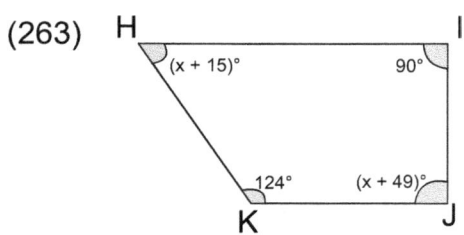

(264)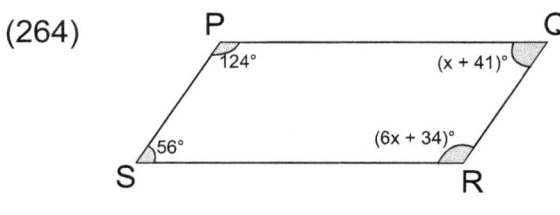

(265)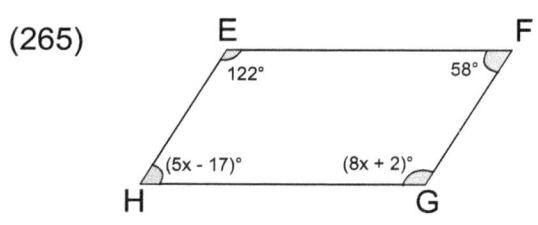

(266)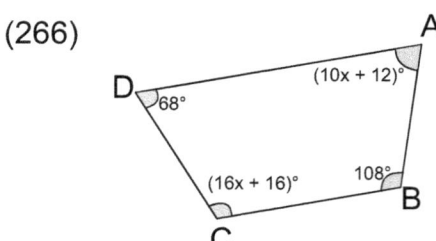

Find the missing value of |x from the below figures.
Remember to use the angle relationships.

(267)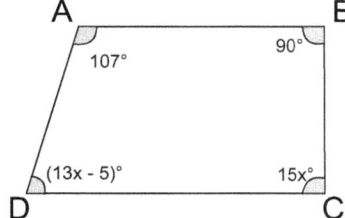

(268)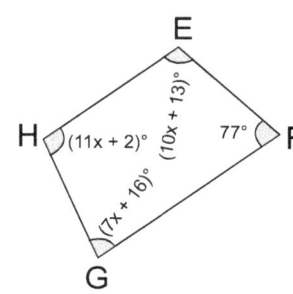

(269)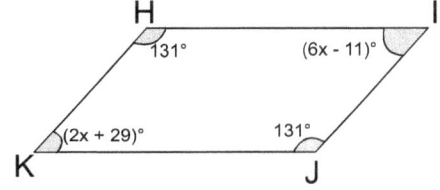

(270)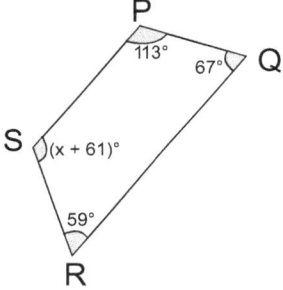

(271)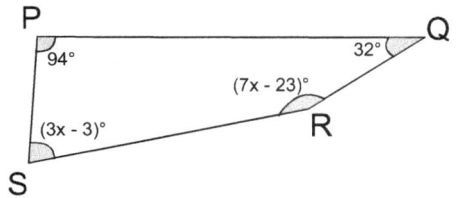

(272)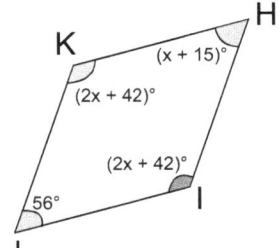

Find the missing value of |x from the below figures.
Remember to use the angle relationships.

(273)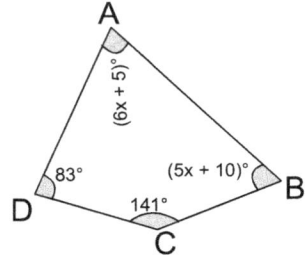

(274)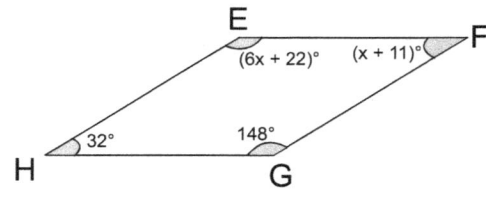

(275)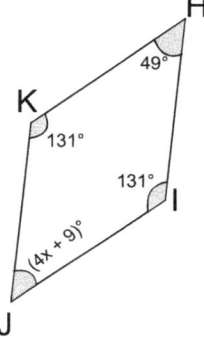

(276)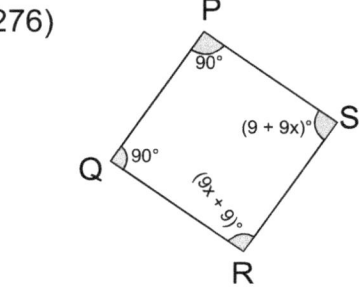

(277)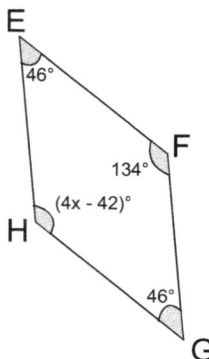

(278)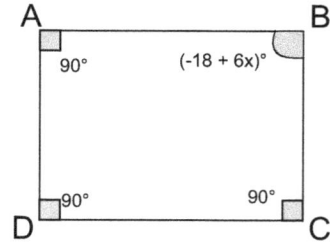

GEOMETRY

Find the missing value of x from the below figures.
Remember to use the angle relationships.

(279)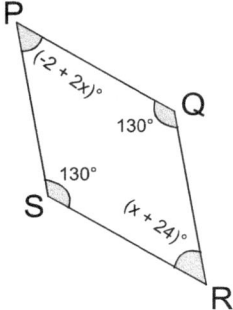

(280)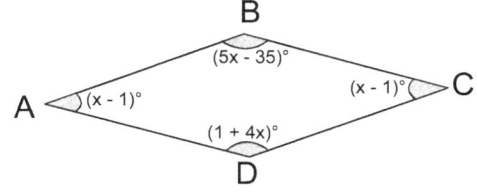

(281)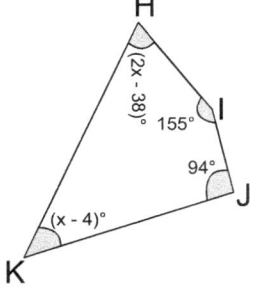

(282)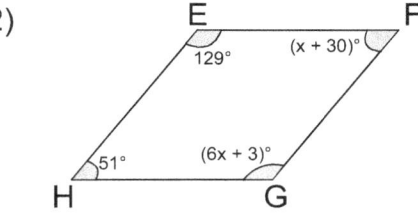

(283)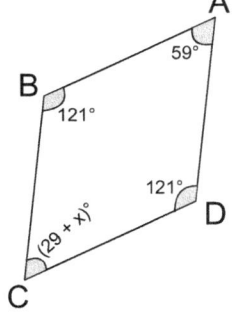

(284)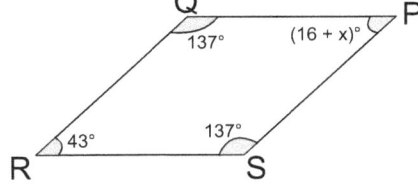

GEOMETRY

Find the missing value of |x from the below figures.
Remember to use the angle relationships.

(285)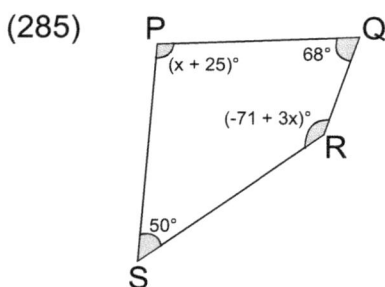

GEOMETRY

Find the area of each figure given below.

(286)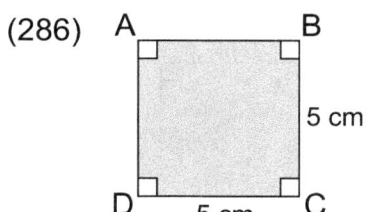

(287)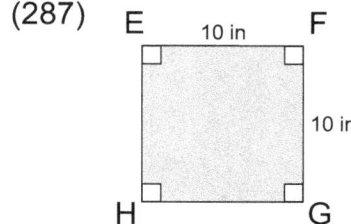

(288)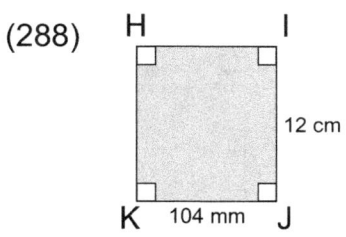

(289)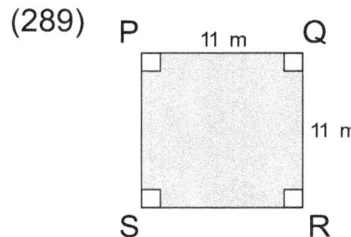

(290)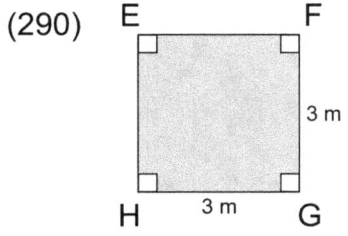

(291)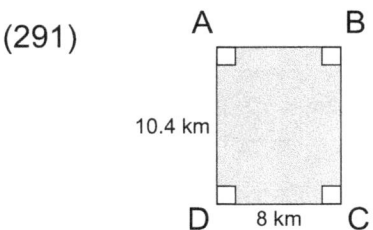

Find the area of each figure given below.

(292)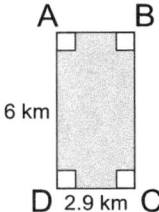
A B
6 km
D 2.9 km C

(293)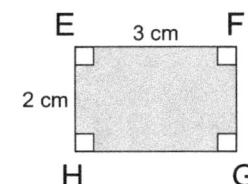
E 3 cm F
2 cm
H G

(294)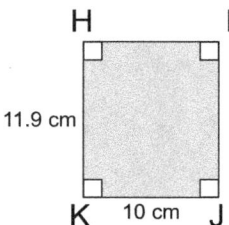
H I
11.9 cm
K 10 cm J

(295)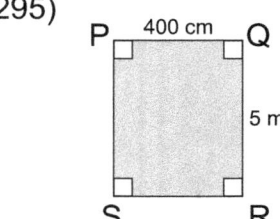
P 400 cm Q
5 m
S R

(296)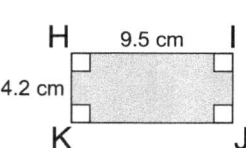
H 9.5 cm I
4.2 cm
K J

(297)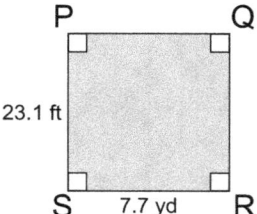
P Q
23.1 ft
S 7.7 yd R

Find the area of each figure given below.

(298)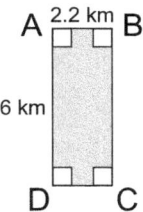

(299)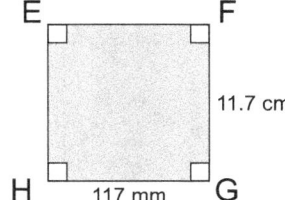

(300)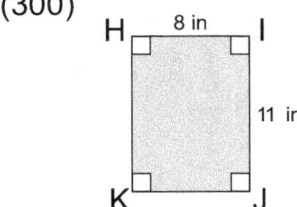

(301)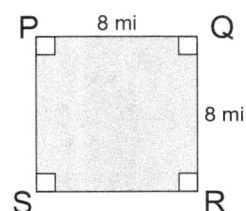

(302)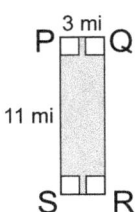

(303)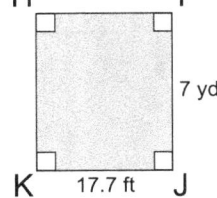

Find the area of each figure given below.

(304)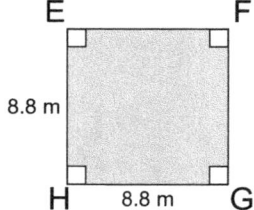

(305)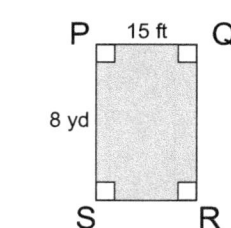

(306)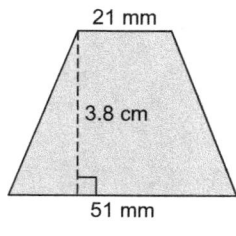

ABCD

(307)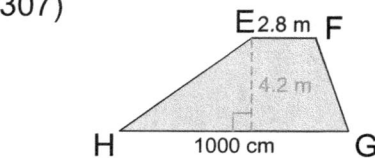

(308)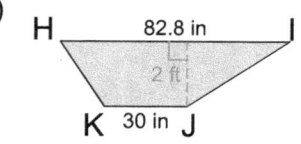

(309)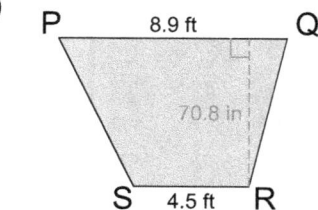

Find the area of each figure given below.

(310)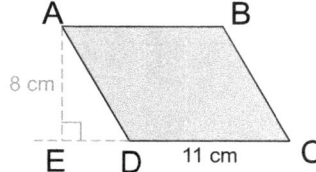

(311)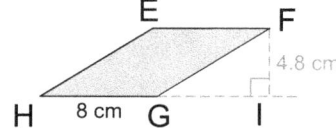

(312)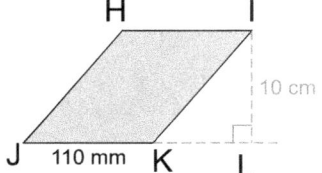

(313)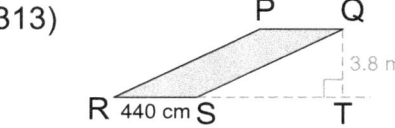

(314)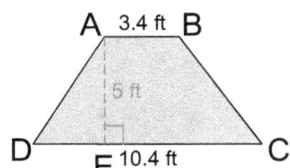

(315)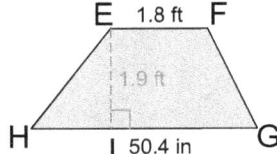

Find the area of each figure given below.

(316)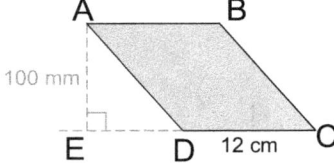

(317)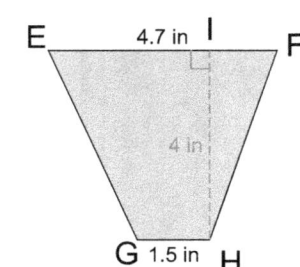

(318)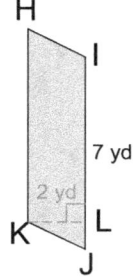

(319)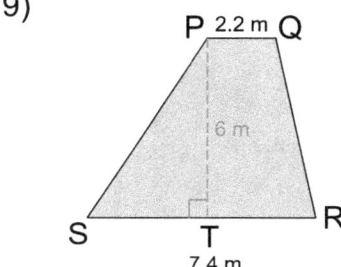

(320)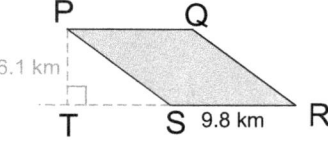

(321)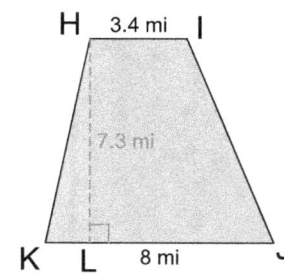

Find the area of each figure given below.

(322)

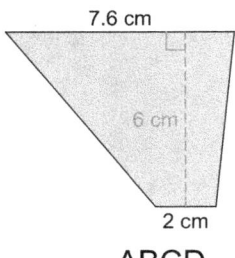

7.6 cm
6 cm
2 cm

ABCD

(323)

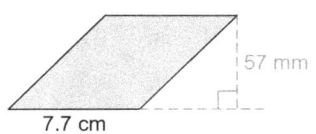

57 mm
7.7 cm

EFGH

(324)

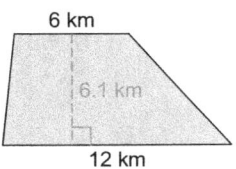

6 km
6.1 km
12 km

HIJK

(325)

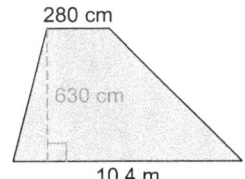

280 cm
630 cm
10.4 m

PQRS

(326) Find the area of a circle with radius as 6 miles.

(327) Find the area of a circle with radius as 3 miles.

(328) Find the area of a circle with radius as 11 miles.

(329) Find the area of a circle with radius as 7 miles.

(330) Find the area of a circle with radius as 9 miles.

Find the area of each of the below figures. (Round your answer to the nearest whole).

(331)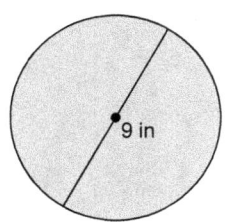

(332)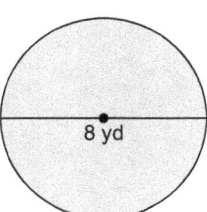

(333)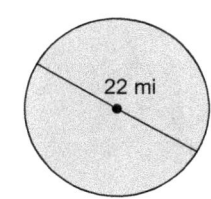

(334)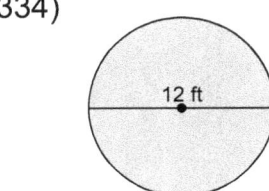

(335)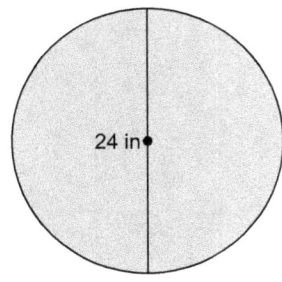

(336)

Find the area of each of the below figures. (Round your answer to the nearest whole).

(337)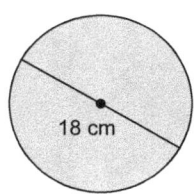

(338)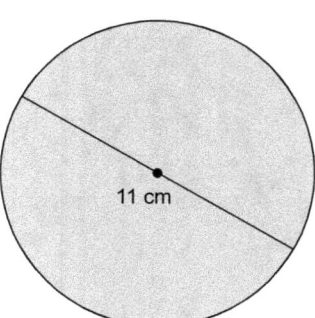

(339)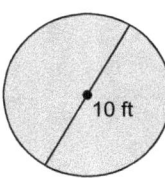

(340)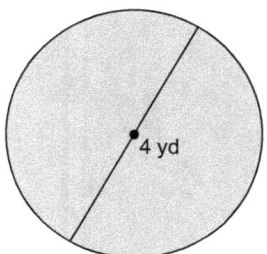

Find the circumference of each of the below figures.
(Round your answer to the nearest whole).

(341)

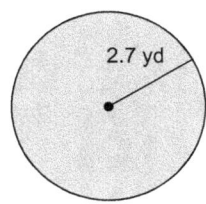

(342)

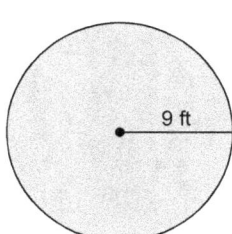

(343)

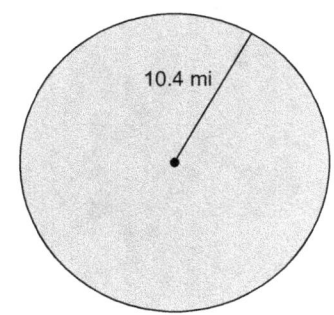

(344)

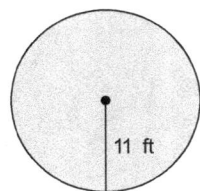

(345)

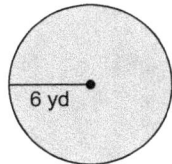

(346)

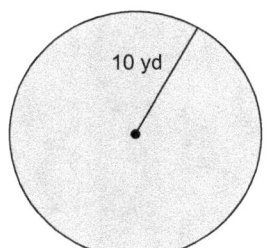

GEOMETRY

Find the circumference of each of the below figures.
(Round your answer to the nearest whole).

(347)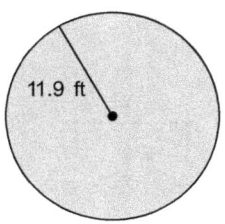

(348)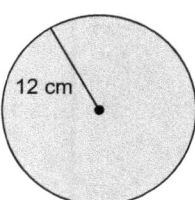

(349)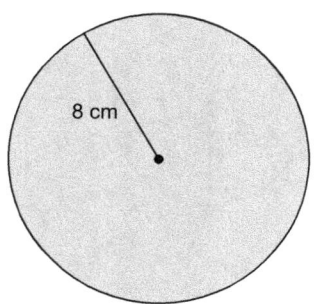

(350)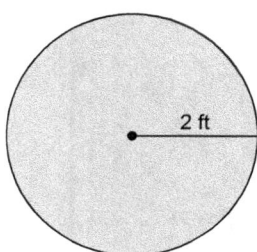

(351)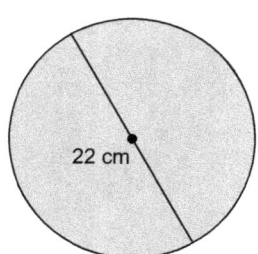

(352)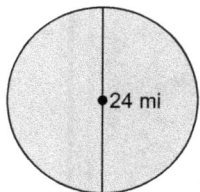

GEOMETRY

Find the circumference of each of the below figures.
(Round your answer to the nearest whole).

(353)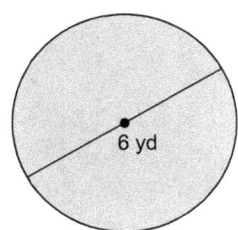

(354)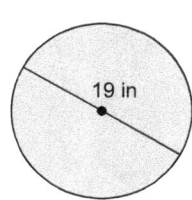

(355)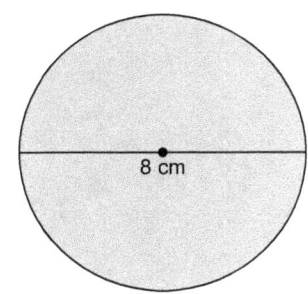

(356)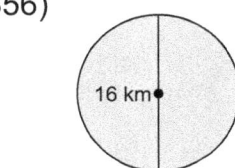

(357)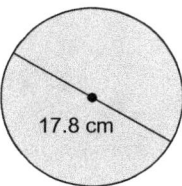

(358)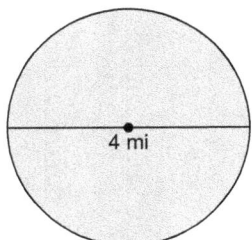

Find the circumference of each of the below figures.
(Round your answer to the nearest whole).

(359)

(360)

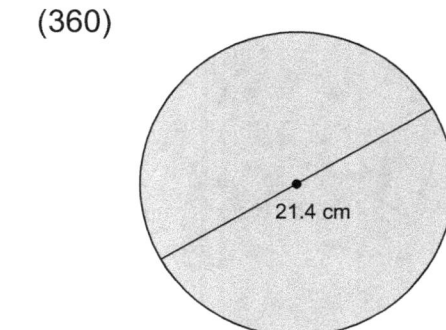

19.2 km

21.4 cm

(361) Find the radius of each circle given the circumference as 37.7 cm

(362) Find the radius of each circle given the circumference as 44 cm

(363) Find the radius of each circle given the circumference as 68.5 mi

(364) Find the radius of each circle given the circumference as 22 mi

(365) Find the radius of each circle given the circumference as 31.4 km

(366) Find the radius of each circle given the circumference as 18.8 yd

(367) Find the radius of each circle given the circumference as 25.1 m

(368) Find the radius of each circle given the circumference as 57.2 km

(369) Find the radius of each circle given the circumference as 75.4 ft

(370) Find the radius of each circle given the circumference as 12.6 cm

(371) Find the diameter of each circle given the circumference as 62.8 mi

(372) Find the diameter of each circle given the circumference as 18.8 km

(373) Find the diameter of each circle given the circumference as 32.7 cm

(374) Find the diameter of each circle given the circumference as 44 km

(375) Find the diameter of each circle given the circumference as 22 m

(376) Find the diameter of each circle given the circumference as 37.1 in

(377) Find the diameter of each circle given the circumference as 56.5 cm

(378) Find the diameter of each circle given the circumference as 75.4 km

(379) Find the diameter of each circle given the circumference as 19.5 cm

(380) Find the diameter of each circle given the circumference as 47.1 ft

(381) Find the area of the circle given circumference as 56.5 km. (Round your answer to the nearest tenth.)

(382) Find the area of the circle given circumference as 12.6 cm. (Round your answer to the nearest tenth.)

(383) Find the area of the circle given circumference as 35.2 cm. (Round your answer to the nearest tenth.)

(384) Find the area of the circle given circumference as 46.5 mi. (Round your answer to the nearest tenth.)

(385) Find the area of the circle given circumference as 25.1 cm. (Round your answer to the nearest tenth.)

(386) Find the area of the circle given circumference as 18.8 m. (Round your answer to the nearest tenth.)

(387) Find the area of the circle given circumference as 75.4 yd. (Round your answer to the nearest tenth.)

(388) Find the area of the circle given circumference as 69.1 m. (Round your answer to the nearest tenth.)

(389) Find the area of the circle given circumference as 37.7 in. (Round your answer to the nearest tenth.)

(390) Find the area of the circle given circumference as 44 cm. (Round your answer to the nearest tenth.)

(391) Find the volume of the sphere with diameter as 4 yd.
(Round your answer to the nearest tenth.)

(392) Find the volume of the cone with radius as 4 cm and a height of 11 cm.
(Round your answer to the nearest tenth.)

(393) Find the volume of the sphere with diameter as 9 cm.
(Round your answer to the nearest tenth.)

(394) Find the volume of the cylinder with radius as 2 in and a height of 5 in.
(Round your answer to the nearest tenth.)

(395) Find the volume of the cylinder with radius as 6 yd and a height of 6 yd.
(Round your answer to the nearest tenth.)

(396) Find the volume of the sphere with radius as 9.9 mi.
(Round your answer to the nearest tenth.)

(397) Find the volume of the cylinder with radius as 2 yd and a height of 7 yd.
(Round your answer to the nearest tenth.)

(398) Find the volume of the cylinder with diameter as 6 cm and a height of 2 cm.
(Round your answer to the nearest tenth.)

(399) Find the volume of the sphere with diameter as 2 ft.
(Round your answer to the nearest tenth.)

(400) Find the volume of the sphere with diameter as 6 cm.
(Round your answer to the nearest tenth.)

(401) Find the volume of the cone with radius as 4 in and a height of 9 in.
(Round your answer to the nearest tenth.)

(402) Find the volume of the cylinder with radius as 12 ft and a height of 10 ft.
(Round your answer to the nearest tenth.)

(403) Find the volume of the cylinder with diameter as 12 km and a height of 11 km.
(Round your answer to the nearest tenth.)

(404) Find the volume of the cylinder with radius as 7 cm and a height of 9 cm.
(Round your answer to the nearest tenth.)

(405) Find the volume of the sphere with radius as 11 in.
(Round your answer to the nearest tenth.)

(406) Find the volume of the cylinder with diameter as 16 mi and a height of 10 mi.
(Round your answer to the nearest tenth.)

(407) Find the volume of the sphere with diameter as 1.4 m.
(Round your answer to the nearest tenth.)

(408) Find the volume of the cylinder with radius as 3 ft and a height of 7 ft.
(Round your answer to the nearest tenth.)

(411) Find the volume of the sphere with diameter as 23.8 ft.
(Round your answer to the nearest tenth.)

(409) Find the volume of the cone with radius as 11 mi and a height of 22 mi.
(Round your answer to the nearest tenth.)

(410) Find the volume of the cylinder with diameter as 16 yd and a height of 16 yd. (Round your answer to the nearest tenth.)

(412) Find the volume of the cylinder with radius as 6 in and a height of 4 in. (Round your answer to the nearest tenth.)

(413) Find the volume of the cone with diameter as 12 in and a height of 24 in. (Round your answer to the nearest tenth.)

(414) Find the volume of the cylinder with radius as 4 mi and a height of 12 mi. (Round your answer to the nearest tenth.)

(415) Find the volume of the cylinder with radius as 8 km and a height of 12 km. (Round your answer to the nearest tenth.)

(416) Find the volume of a rectangular prism measuring 14 yd and 20 yd along the base and 9 yd tall. (Round your answer to the nearest tenth.)

(417) Find the volume of a rectangular prism measuring 10 mi and 14 mi along the base and 9 mi tall. (Round your answer to the nearest tenth.)

(418) Find the volume of a rectangular prism measuring 20 km and 10 km along the base and 8 km tall. (Round your answer to the nearest tenth.)

(419) Find the volume of a rectangular prism measuring 18 m and 15 m along the base and 6 m tall. (Round your answer to the nearest tenth.)

(420) Find the volume of a square prism measuring 19 km along each edge of the base and 20 km tall. (Round your answer to the nearest tenth.)

(421) Find the volume of a rectangular prism measuring 9 ft and 16 ft along the base and 12 ft tall. (Round your answer to the nearest tenth.)

(422) Find the volume of a rectangular prism measuring 13 yd and 10 yd along the base and 18 yd tall. (Round your answer to the nearest tenth.)

(423) Find the volume of a rectangular prism measuring 15 m and 20 m along the base and 10 m tall. (Round your answer to the nearest tenth.)

(424) Find the volume of a square prism measuring 3 cm along each edge of the base and 13 cm tall. (Round your answer to the nearest tenth.)

(425) Find the volume of a square prism measuring 3 yd along each edge of the base and 10 yd tall. (Round your answer to the nearest tenth.)

(426) Find the volume of a square prism measuring 10 cm along each edge of the base and 3 cm tall. (Round your answer to the nearest tenth.)

(427) Find the volume of a square prism measuring 16 m along each edge of the base and 5 m tall. (Round your answer to the nearest tenth.)

(428) Find the volume of a rectangular prism measuring 16 in and 2 in along the base and 7 in tall. (Round your answer to the nearest tenth.)

(429) Find the volume of a rectangular prism measuring 14 in and 11 in along the base and 5 in tall. (Round your answer to the nearest tenth.)

(430) Find the volume of a rectangular prism measuring 7 yd and 9 yd along the base and 16 yd tall. (Round your answer to the nearest tenth.)

(431) Find the volume of a rectangular prism measuring 13 yd and 16 yd along the base and 18 yd tall. (Round your answer to the nearest tenth.)

(432) Find the volume of a rectangular prism measuring 6 mi and 3 mi along the base and 3 mi tall. (Round your answer to the nearest tenth.)

(433) Find the volume of a square prism measuring 18 km along each edge of the base and 14 km tall. (Round your answer to the nearest tenth.)

(434) Find the volume of a rectangular prism measuring 14 m and 5 m along the base and 5 m tall. (Round your answer to the nearest tenth.)

(435) Find the volume of a square prism measuring 9 yd along each edge of the base and 9 yd tall. (Round your answer to the nearest tenth.)

(436) Find the volume of a square prism measuring 15 ft along each edge of the base and 10 ft tall. (Round your answer to the nearest tenth.)

(437) Find the volume of a square prism measuring 11 in along each edge of the base and 7 in tall. (Round your answer to the nearest tenth.)

(438) Find the volume of a rectangular prism measuring 20 m and 10 m along the base and 20 m tall. (Round your answer to the nearest tenth.)

(439) Find the volume of a square prism measuring 9 yd along each edge of the base and 2 yd tall. (Round your answer to the nearest tenth.)

(440) Find the volume of a rectangular prism measuring 11 mi and 13 mi along the base and 12 mi tall. (Round your answer to the nearest tenth.)

Find the volume of each of the below figures. (Round your answer to the nearest whole).

(441)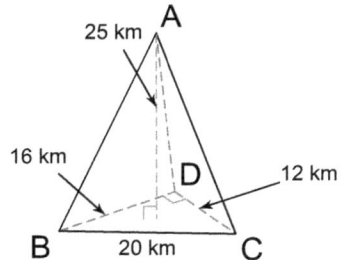

(442)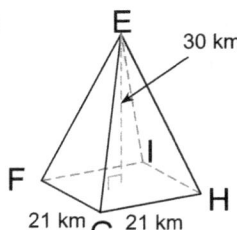

(443)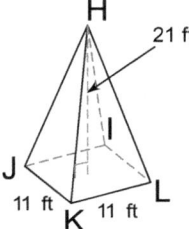

(444)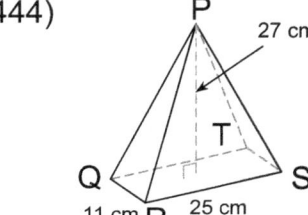

(445)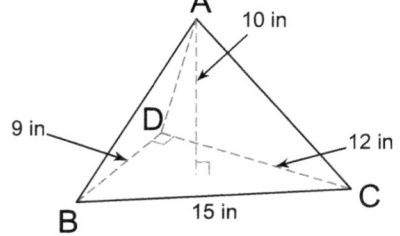

(446)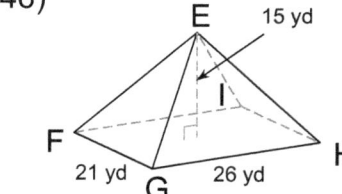

Find the volume of each of the below figures. (Round your answer to the nearest whole).

(447)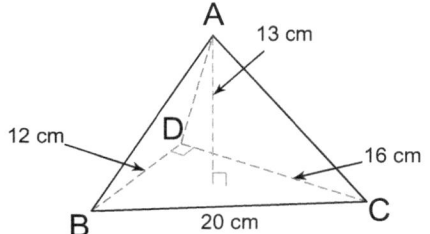

(448)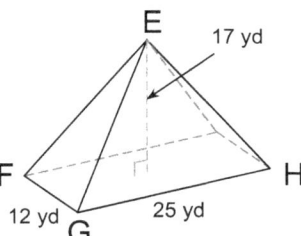

(449)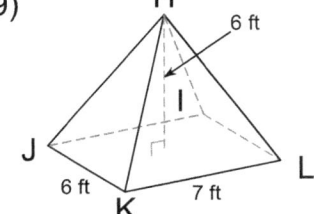

(450)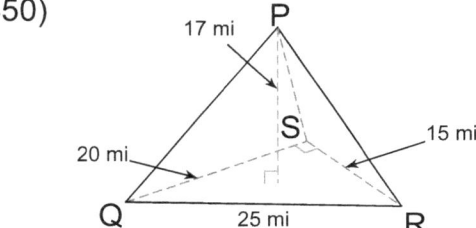

(451)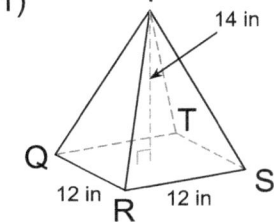

(452)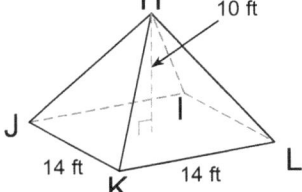

Find the volume of each of the below figures. (Round your answer to the nearest whole).

(453)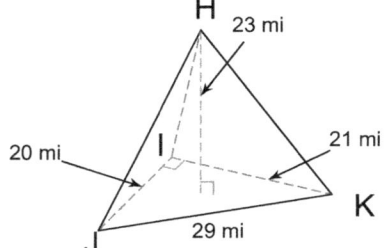

(454)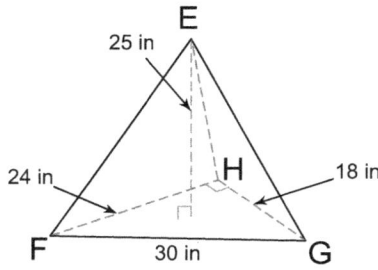

(455)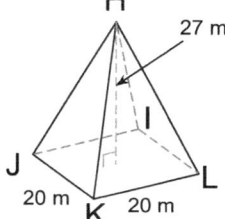

(456)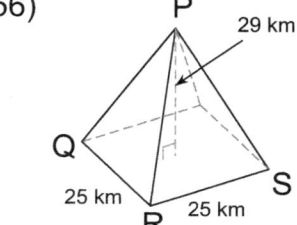

(457)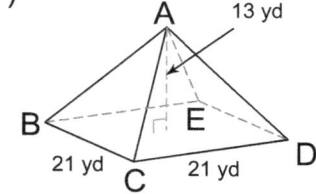

(458)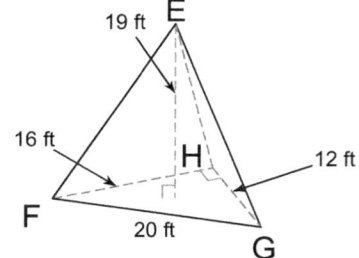

Find the volume of each of the below figures. (Round your answer to the nearest whole).

(459)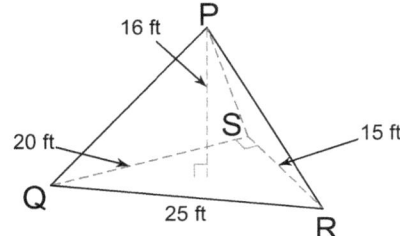

(460)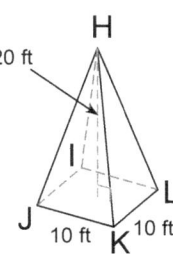

(461)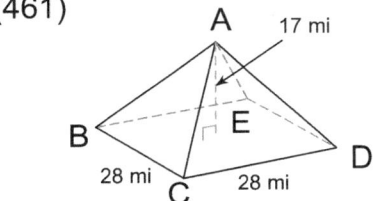

(462)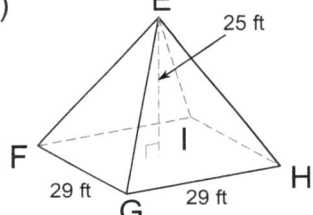

(463)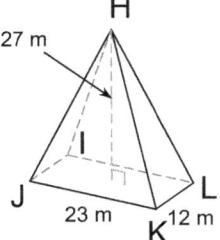

(464)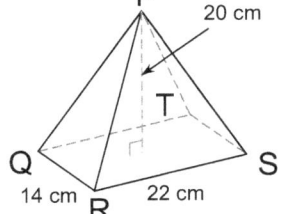

Find the volume of each of the below figures. (Round your answer to the nearest whole).

(465)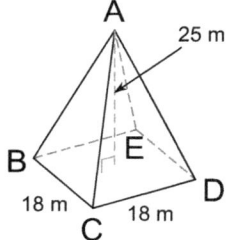

(466)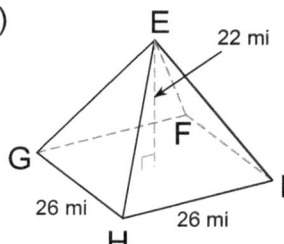

(467)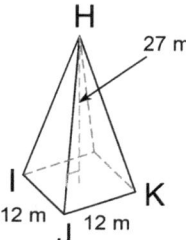

(468)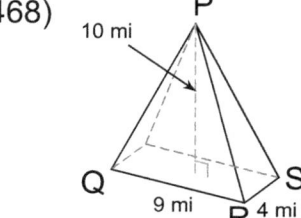

(469)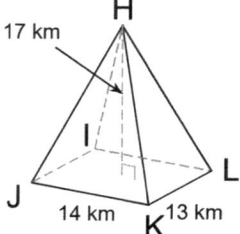

(470)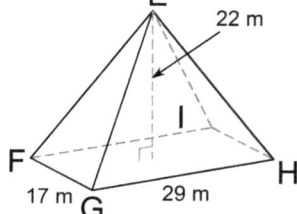

(471) Find the total surface area of a cylinder with a radius of 14 yd and a height of 14 yd. (Round to the nearest tenth.)

(472) Find the total surface area of a sphere with a radius of 3 in. (Round to the nearest tenth.)

(473) Find the total surface area of a cone with radius 9 ft and a slant height of 21 ft. (Round to the nearest tenth.)

(474) Find the total surface area of a cylinder with a radius of 12 ft and a height of 6 ft. (Round to the nearest tenth.)

(475) Find the total surface area of a cone with diameter 14 m and a slant height of 15.7 m. (Round to the nearest tenth.)

(476) Find the total surface area of a cone with diameter 22 mi and a slant height of 24.6 mi. (Round to the nearest tenth.)

(477) Find the total surface area of a sphere with a diameter of 34 in. (Round to the nearest tenth.)

(478) Find the total surface area of a sphere with a radius of 10 km. (Round to the nearest tenth.)

(479) Find the total surface area of a sphere with a radius of 8.9 in. (Round to the nearest tenth.)

(480) Find the total surface area of a cylinder with a diameter of 38 km and a height of 14 km. (Round to the nearest tenth.)

(481) Find the total surface area of a cylinder with a radius of 6 km and a height of 20 km. (Round to the nearest tenth.)

(482) Find the total surface area of a cone with radius 4 in and a slant height of 18.4 in. (Round to the nearest tenth.)

(483) Find the total surface area of a cylinder with a diameter of 16 in and a height of 16 in. (Round to the nearest tenth.)

(484) Find the total surface area of a cylinder with a diameter of 18 cm and a height of 16 cm. (Round to the nearest tenth.)

(485) Find the total surface area of a cylinder with a diameter of 6 m and a height of 3 m. (Round to the nearest tenth.)

(486) Find the total surface area of a sphere with a radius of 9 yd. (Round to the nearest tenth.)

(487) Find the total surface area of a cone with diameter 2 ft and a slant height of 12 ft. (Round to the nearest tenth.)

(488) Find the total surface area of a cone with radius 16 ft and a slant height of 35.8 ft. (Round to the nearest tenth.)

(489) Find the total surface area of a cone with diameter 26 ft and a slant height of 29.1 ft. (Round to the nearest tenth.)

(490) Find the total surface area of a cone with diameter 28 mi and a slant height of 31.3 mi. (Round to the nearest tenth.)

(491) Find the total surface area of a cylinder with a diameter of 36 cm and a height of 11 cm. (Round to the nearest tenth.)

(492) Find the total surface area of a cylinder with a diameter of 4 yd and a height of 1 yd. (Round to the nearest tenth.)

(493) Find the total surface area of a cylinder with a radius of 7 m and a height of 5 m. (Round to the nearest tenth.)

(494) Find the total surface area of a cylinder with a diameter of 36 cm and a height of 17 cm. (Round to the nearest tenth.)

(495) Find the total surface area of a cone with radius 12 in and a slant height of 26.8 in. (Round to the nearest tenth.)

(496) Find the total surface area of a cone with radius 1 in and a slant height of 10 in. (Round to the nearest tenth.)

(497) Find the total surface area of a cone with diameter 18 km and a slant height of 21.9 km. (Round to the nearest tenth.)

(498) Find the total surface area of a sphere with a diameter of 30 cm. (Round to the nearest tenth.)

(499) Find the total surface area of a cylinder with a diameter of 24 km and a height of 17 km. (Round to the nearest tenth.)

(500) Find the total surface area of a cylinder with a diameter of 4 mi and a height of 9 mi. (Round to the nearest tenth.)

(501) Find the total surface area of a square prism measuring 13 yd along each edge of the base and 9 yd tall. (Round to the nearest tenth.)

(502) Find the total surface area of a rectangular prism measuring 14 mi and 19 mi along the base and 19 mi tall. (Round to the nearest tenth.)

(503) Find the total surface area of a square prism measuring 6 ft along each edge of the base and 8 ft tall. (Round to the nearest tenth.)

(504) Find the total surface area of a square prism measuring 18 mi along each edge of the base and 10 mi tall. (Round to the nearest tenth.)

(505) Find the total surface area of a rectangular prism measuring 18 ft and 20 ft along the base and 16 ft tall. (Round to the nearest tenth.)

(506) Find the total surface area of a square prism measuring 7 in along each edge of the base and 2 in tall. (Round to the nearest tenth.)

(507) Find the total surface area of a rectangular prism measuring 17 ft and 19 ft along the base and 14 ft tall. (Round to the nearest tenth.)

(508) Find the total surface area of a square pyramid measuring 16 cm along the base with a slant height of 12.8 cm. (Round to the nearest tenth.)

(509) Find the total surface area of a rectangular prism measuring 12 m and 6 m along the base and 20 m tall. (Round to the nearest tenth.)

(510) Find the total surface area of a rectangular prism measuring 11 in and 10 in along the base and 17 in tall. (Round to the nearest tenth.)

(511) Find the total surface area of a square prism measuring 15 m along each edge of the base and 9 m tall. (Round to the nearest tenth.)

(512) Find the total surface area of a rectangular prism measuring 10 mi and 12 mi along the base and 2 mi tall. (Round to the nearest tenth.)

(513) Find the total surface area of a square pyramid measuring 4 km along the base with a slant height of 6.3 km. (Round to the nearest tenth.)

(514) Find the total surface area of a square prism measuring 18 cm along each edge of the base and 12 cm tall. (Round to the nearest tenth.)

(515) Find the total surface area of a square prism measuring 5 m along each edge of the base and 5 m tall. (Round to the nearest tenth.)

(516) Find the total surface area of a square pyramid measuring 15 km along the base with a slant height of 14.2 km. (Round to the nearest tenth.)

(517) Find the total surface area of a rectangular prism measuring 7 cm and 5 cm along the base and 3 cm tall. (Round to the nearest tenth.)

(518) Find the total surface area of a rectangular prism measuring 8 km and 7 km along the base and 11 km tall. (Round to the nearest tenth.)

(519) Find the total surface area of a square pyramid measuring 7 cm along the base with a slant height of 8.7 cm. (Round to the nearest tenth.)

(520) Find the total surface area of a square prism measuring 18 yd along each edge of the base and 15 yd tall. (Round to the nearest tenth.)

(521) Find the total surface area of a square prism measuring 2 m along each edge of the base and 5 m tall. (Round to the nearest tenth.)

(522) Find the total surface area of a square prism measuring 10 mi along each edge of the base and 18 mi tall. (Round to the nearest tenth.)

(523) Find the total surface area of a rectangular prism measuring 11 cm and 16 cm along the base and 3 cm tall. (Round to the nearest tenth.)

(524) Find the total surface area of a rectangular prism measuring 7 m and 13 m along the base and 2 m tall. (Round to the nearest tenth.)

(525) Find the total surface area of a square prism measuring 12 in along each edge of the base and 5 in tall. (Round to the nearest tenth.)

(526) Find the total surface area of a square prism measuring 6 ft along each edge of the base and 6 ft tall. (Round to the nearest tenth.)

(527) Find the total surface area of a square prism measuring 12 cm along each edge of the base and 10 cm tall. (Round to the nearest tenth.)

(528) Find the total surface area of a rectangular prism measuring 4 cm and 7 cm along the base and 13 cm tall. (Round to the nearest tenth.)

(529) Find the total surface area of a square pyramid measuring 6 mi along the base with a slant height of 11.4 mi. (Round to the nearest tenth.)

(530) Find the total surface area of a square pyramid measuring 13 km along the base with a slant height of 14.5 km. (Round to the nearest tenth.)

Find the total surface area of the below figures.
(Round to the nearest tenth.)

(531)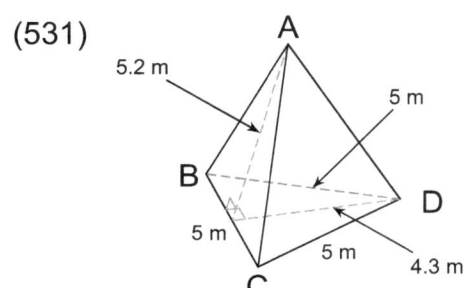

(532)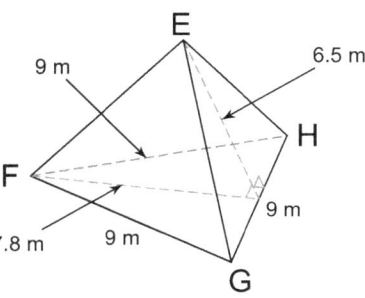

(533)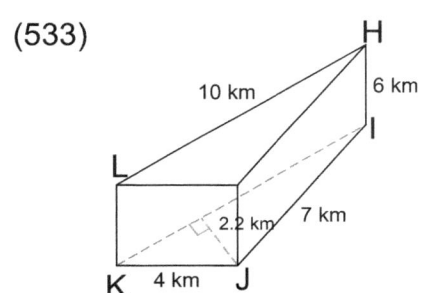

(534)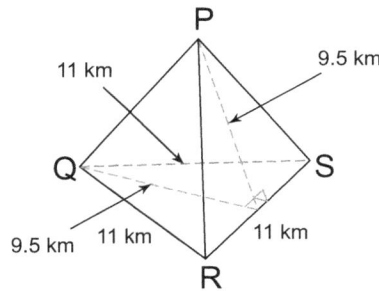

(535)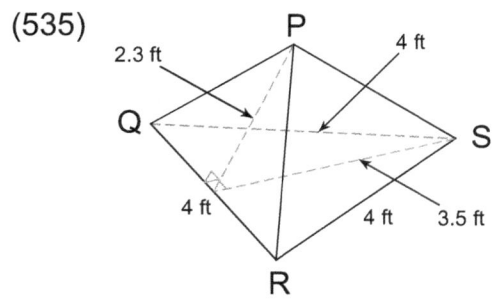

(536)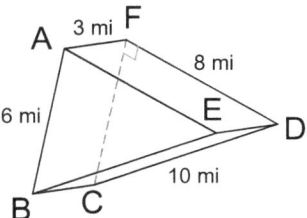

Find the total surface area of the below figures.
(Round to the nearest tenth.)

(537)
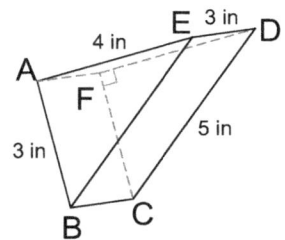

(538)
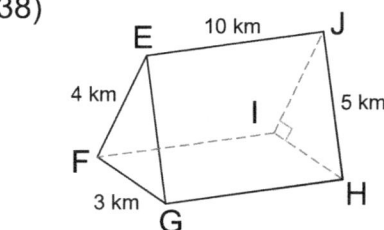

(539)
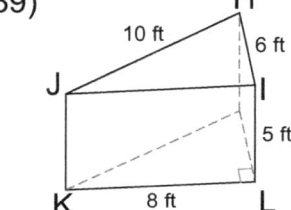

(540)
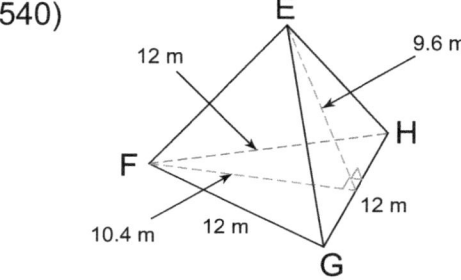

(541)
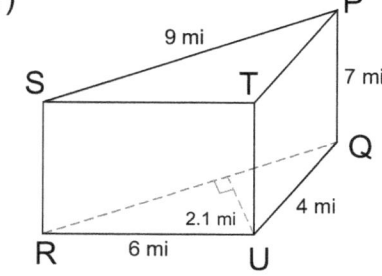

(542)
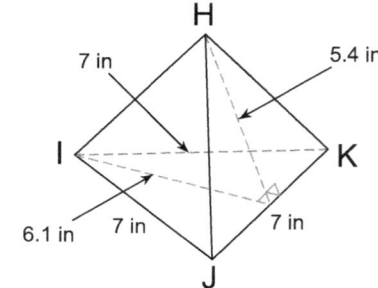

Find the total surface area of the below figures.
(Round to the nearest tenth.)

(543)

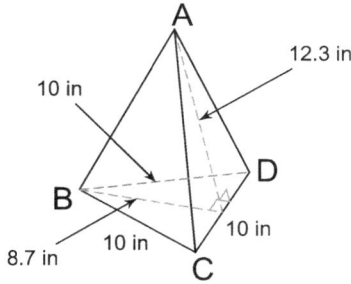

(544)

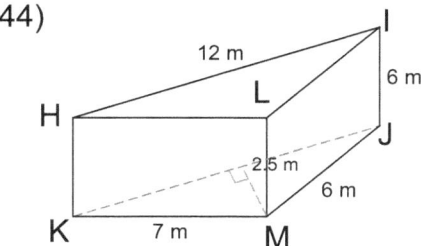

(545)

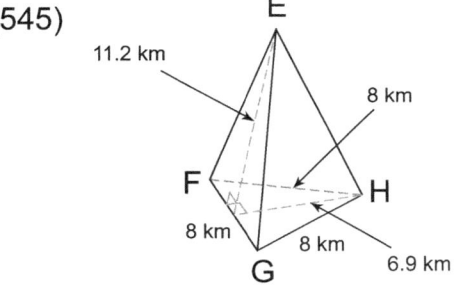

(546)

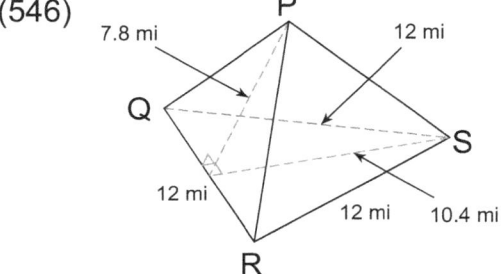

(547)

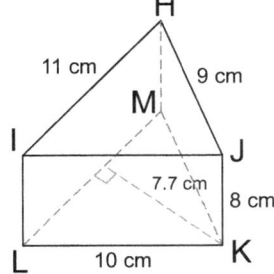

(548)

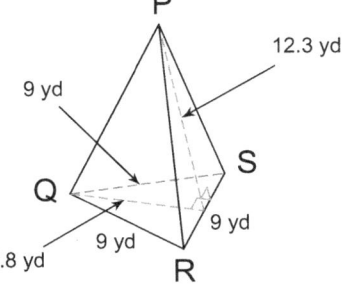

Find the total surface area of the below figures.
(Round to the nearest tenth.)

(549)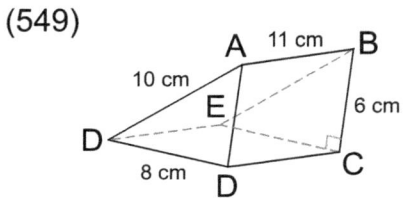

(550)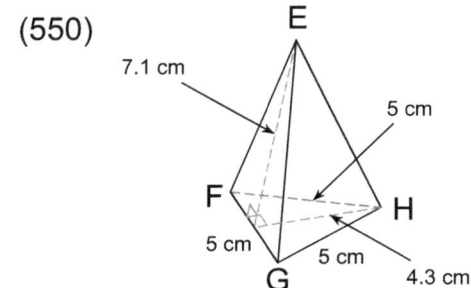

(1) alternate exterior (2) alternate interior

(3) alternate interior (4) corresponding

(5) supplementary (6) alternate interior

(7) vertical (8) adjacent

(9) complementary (10) vertical

(11) alternate exterior (12) complementary

(13) complementary (14) complementary

(15) vertical   (16) complementary

(17) complementary   (18) corresponding

(19) alternate exterior   (20) complementary

(21) corresponding   (22) corresponding

(23) alternate interior   (24) supplementary

(25) alternate exterior   (26) adjacent

(27) supplementary   (28) corresponding

GEOMETRY

Basic Math Answer Keys

(29) alternate interior (30) vertical

(31) 26° (32) 136°

(33) 60° (34) 114°

(35) 139° (36) 93°

(37) 114° (38) 102°

(39) 78° (40) 55°

(41) 98° (42) 86°

GEOMETRY

Basic Math Answer Keys

(43) 25°   (44) 30°

(45) 49°   (46) 143°

(47) 137°   (48) 65°

(49) 33°   (50) 87°

(51) 29°   (52) 89°

(53) 44°   (54) 36°

(55) 72°   (56) 39°

GEOMETRY

Basic Math Answer Keys

(57)   104°   (58)   34°

(59)   72°   (60)   24°

(61)   70°   (62)   31°

(63)   91°   (64)   45°

(65)   38°   (66)   59°

(67)   87°   (68)   42°

(69)   31°   (70)   77°

GEOMETRY

Basic Math Answer Keys

(71) 84°   (72) 124°

(73) 52°   (74) 90°

(75) 34°   (76) 58°

(77) 72°   (78) 65°

(79) 85°   (80) 87°

(81) 65°   (82) 61°

(83) 81°   (84) 95°

GEOMETRY

Basic Math Answer Keys

(85) 146°   (86) 74°

(87) 46°   (88) 78°

(89) 38°   (90) 55°

(91) 52°   (92) 41°

(93) 32°   (94) 59°

(95) 78°   (96) 77°

(97) 102°   (98) 93°

GEOMETRY

Basic Math Answer Keys

(99)   59°   (100)   113°

(101) 8   (102)   12

(103) 16   (104)   15

(105) 67   (106)   34

(107) 15   (108)   28

(109) 39   (110)   39

(111) 38   (112)   20

Basic Math
Answer Keys

(113) 17   (114) 11

(115) 16   (116) 65

(117) 22   (118) 72

(119) 48   (120) 12

(121) 50   (122) 7

(123) 59   (124) 4

(125) 58   (126) 24

GEOMETRY

(127) 98 (128) 29

(129) 11 (130) 42

(131) 30 (132) 6

(133) 20 (134) 49

(135) 9   (136) 22

(137) 25 (138) 23

(139) 4   (140) 26

GEOMETRY

Basic Math Answer Keys

(141) 7   (142) 3

(143) 73   (144) 3

(145) 16   (146) 5

(147) 5   (148) 8

(149) 5   (150) 22

(151) 8   (152) 19

(153) 16   (154) 3

GEOMETRY

Basic Math Answer Keys

(155) 3   (156) 8

(157) 6   (158) 11

(159) 9   (160) 7

(161) 11   (162) 12

(163) 3   (164) 3

(165) 11   (166) 4

(167) 9   (168) 7

Basic Math Answer Keys

(169) 13

(170) 7

(171) 81°

(172) 114°

(173) 65°

(174) 59°

(175) 42°

(176) 58°

(177) 29°

(178) 45°

(179) 21°

(180) 25°

(181) 24°

(182) 119°

(183) 92°   (184) 54°

(185) 99°   (186) 50°

(187) 60°   (188) 47°

(189) 105°   (190) 45°

(191) 96°   (192) 40°

(193) 37°   (194) 69°

(195) 45°   (196) 25°

 GEOMETRY

(197) 126°   (198) 75°

(199) 37°   (200) 79°

(201) 15   (202) 2

(203) 5   (204) 3

(205) 8   (206) 4

(207) 13   (208) 8

(209) 6   (210) 13

GEOMETRY

Basic Math Answer Keys

(211) 15   (212) 5

(213) 4   (214) 9

(215) 26   (216) 22

(217) 8   (218) 5

(219) 9   (220) 3

(221) 11   (222) 17

(223) 4   (224) 8

(225) 8   (226) 8

(227) 9   (228) 9

(229) 8   (230) 8

(231) 33   (232) 15

(233) 67   (234) 73

(235) 29   (236) 14

(237) 12   (238) 8

(239) 35   (240) 32

(241) 22   (242) 13

(243) 12   (244) 18

(245) 16   (246) 15

(247) 42   (248) 36

(249) 22   (250) 34

(251) 16   (252) 6

GEOMETRY

Basic Math Answer Keys

(253)   8   (254)   19

(255)   7   (256)   7

(257)   25   (258)   18

(259)   16   (260)   7

(261)   14   (262)   55

(263)   41   (264)   15

(265)   15   (266)   6

GEOMETRY

Basic Math Answer Keys

(267) 6   (268) 9

(269) 10   (270) 60

(271) 26   (272) 41

(273) 11   (274) 21

(275) 10   (276) 9

(277) 44   (278) 18

(279) 26   (280) 36

GEOMETRY

Basic Math Answer Keys

(281) 51

(282) 21

(283) 30

(284) 27

(285) 72

(286) 25 cm²

(287) 100 in²

(288) 124.8 cm²

(289) 121 m²

(290) 9 m²

(291) 83.2 km²

(292) 17.4 km²

(293) 6 cm²

(294) 119 cm²

(295) 20 m² (296) 39.9 cm²

(297) 59.29 yd² (298) 13.2 km²

(299) 136.89 cm² (300) 88 in²

(301) 64 mi² (302) 33 mi²

(303) 41.3 yd² (304) 77.44 m²

(305) 40 yd² (306) 13.68 cm²

(307) 26.88 m² (308) 9.4 ft²

(309) 39.53 ft²　　　　　　　　(310) 88 cm²

(311) 38.4 cm²　　　　　　　　(312) 110 cm²

(313) 16.72 m²　　　　　　　　(314) 34.5 ft²

(315) 5.7 ft²　　　　　　　　　(316) 120 cm²

(317) 12.4 in²　　　　　　　　(318) 14 yd²

(319) 28.8 m²　　　　　　　　 (320) 59.78 km²

(321) 41.61 mi²　　　　　　　 (322) 28.8 cm²

GEOMETRY

Basic Math Answer Keys

(323) 43.89 cm²   (324) 54.9 km²

(325) 41.58 m²   (326) 36π mi²

(327) 9π mi²   (328) 121π m²

(329) 49π mi²   (330) 81π m²

(331) 64 in²   (332) 50 yd²

(333) 380 mi²   (334) 113 ft²

(335) 452 in²   (336) 284 ft²

GEOMETRY

Basic Math Answer Keys

(337) 254 cm² (338) 95 cm²

(339) 79 ft² (340) 13 yd²

(341) 17 yd (342) 56.5 ft

(343) 65.3 mi (344) 69.1 ft

(345) 37.7 yd (346) 62.8 yd

(347) 74.8 ft (348) 75.4 cm

(349) 50.3 cm (350) 12.6 ft

GEOMETRY

Basic Math Answer Keys

(351) 69.1 cm   (352) 75.4 mi

(353) 18.8 yd   (354) 59.7 in

(355) 25.1 cm   (356) 50.3 km

(357) 55.9 cm   (358) 12.6 mi

(359) 60.3 km   (360) 67.2 cm

(361) 6 cm   (362) 7 cm

(363) 10.9 mi   (364) 3.5 mi

GEOMETRY

Basic Math Answer Keys

(365) 5 km

(366) 3 yd

(367) 4 m

(368) 9.1 km

(369) 12 ft

(370) 2 cm

(371) 20 mi

(372) 6 km

(373) 10.4 cm

(374) 14 km

(375) 7 m

(376) 11.8 in

(377) 18 cm

(378) 24 km

(379) 6.2 cm

(380) 15 ft

(381) 254 km²

(382) 12.6 cm²

(383) 98.6 cm²

(384) 172.1 mi²

(385) 50.1 cm²

(386) 28.1 m²

(387) 452.4 yd²

(388) 380 m²

(389) 113.1 in²

(390) 154.1 cm²

(391) 33.5 yd³

(392) 184.3 cm³

GEOMETRY

Basic Math
Answer Keys

(393) 381.7 cm³   (394) 62.8 in³

(395) 678.6 yd³   (396) 4064.4 mi³

(397) 88 yd³   (398) 56.5 cm³

(399) 4.2 ft³   (400) 113.1 cm³

(401) 150.8 in³   (402) 4523.9 ft³

(403) 1244.1 km³   (404) 1385.4 cm³

(405) 5575.3 in³   (406) 2010.6 mi³

GEOMETRY

Basic Math Answer Keys

(407) 1.4 m³   (408) 197.9 ft³

(409) 2787.6 mi³   (410) 1206.4 yd³

(411) 7058.8 ft³   (412) 452.4 in³

(413) 3619.1 in³   (414) 603.2 mi³

(415) 2412.7 km³   (416) 2520 yd³

(417) 1260 mi³   (418) 1600 km³

(419) 1620 m³   (420) 7220 km³

(421) 1728 ft³   (422) 2340 yd³

(423) 3000 m³   (424) 117 cm³

(425) 90 yd³   (426) 300 cm³

(427) 1280 m³   (428) 224 in³

(429) 770 in³   (430) 1008 yd³

(431) 3744 yd³   (432) 54 mi³

(433) 4536 km³   (434) 350 m³

(435) 729 yd³ (436) 2250 ft³

(437) 847 in³ (438) 4000 m³

(439) 162 yd³ (440) 1716 mi³

(441) 800 km³ (442) 4410 km³

(443) 847 ft³ (444) 2475 cm³

(445) 180 in³ (446) 2730 yd³

(447) 416 cm³ (448) 1700 yd³

(449) 84 ft³

(450) 850 mi³

(451) 672 in³

(452) 653.3 ft³

(453) 1610 mi³

(454) 1800 in³

(455) 3600 m³

(456) 6041.7 km³

(457) 1911 yd³

(458) 608 ft³

(459) 800 ft³

(460) 666.7 ft³

(461) 4442.7 mi³

(462) 7008.3 ft³

GEOMETRY

Basic Math Answer Keys

(463) 2484 m³   (464) 2053.3 cm³

(465) 2700 m³   (466) 4957.3 mi³

(467) 1296 m³   (468) 120 mi³

(469) 1031.3 km³   (470) 3615.3 m³

(471) 2463 yd²   (472) 113.1 in²

(473) 848.2 ft²   (474) 1357.2 ft²

(475) 499.2 m²   (476) 1230.2 mi²

GEOMETRY

(477) 3631.7 in²   (478) 1256.6 km²

(479) 995.4 in²   (480) 3939.6 km²

(481) 980.2 km²   (482) 281.5 in²

(483) 1206.4 in²   (484) 1413.7 cm²

(485) 113.1 m²   (486) 1017.9 yd²

(487) 40.8 ft²   (488) 2603.8 ft²

(489) 1719.4 ft²   (490) 1992.4 mi²

GEOMETRY

Basic Math Answer Keys

(491) 3279.8 cm²   (492) 37.7 yd²

(493) 527.8 m²   (494) 3958.4 cm²

(495) 1462.7 in²   (496) 34.6 in²

(497) 873.7 km²   (498) 2827.4 cm²

(499) 2186.5 km²   (500) 138.2 mi²

(501) 806 yd²   (502) 1786 mi²

(503) 264 ft²   (504) 1368 mi²

GEOMETRY

Basic Math Answer Keys

(505) 1936 ft²   (506) 154 in²

(507) 1654 ft²   (508) 665.6 cm²

(509) 864 m²   (510) 934 in²

(511) 990 m²   (512) 328 mi²

(513) 66.4 km²   (514) 1512 cm²

(515) 150 m²   (516) 651 km²

(517) 142 cm²   (518) 442 km²

GEOMETRY

Basic Math Answer Keys

(519) 170.8 cm²   (520) 1728 yd²

(521) 48 m²   (522) 920 mi²

(523) 514 cm²   (524) 262 m²

(525) 528 in²   (526) 216 ft²

(527) 768 cm²   (528) 342 cm²

(529) 172.8 mi²   (530) 546 km²

(531) 49.8 m²   (532) 122.9 m²

Basic Math Answer Keys

(533) 148 km²   (534) 209 km²

(535) 20.8 ft²   (536) 120 mi²

(537) 48 in²   (538) 132 km²

(539) 168 ft²   (540) 235.2 m²

(541) 151.9 mi²   (542) 78.1 in²

(543) 228 in²   (544) 180 m²

(545) 162 km²   (546) 202.8 mi²

(547) 324.7 cm²

(548) 201.2 yd²

(549) 312 cm²

(550) 64 cm²

www.ingramcontent.com/pod-product-compliance
Lightning Source LLC
Chambersburg PA
CBHW081745100526
44592CB00015B/2308